Unleashing Mobile App Innovation

**Mastering Mobile App Development:
Advanced Techniques and Best Practices**

Nathanial Morrison

Table of Contents

INTRODUCTION

In the ever-evolving landscape of technology, mobile applications have revolutionized how we live, work, and interact. With the ability to connect us to the outside world and simplify daily tasks, mobile apps have become a vital component of our digital lives. With millions of apps available on different platforms, there is more competition than ever to make outstanding mobile experiences. Welcome to "Unleashing Mobile App Innovation: Mastering Mobile App Development - Advanced Techniques and Best Practices," the starting point to unlocking the secrets of crafting outstanding mobile applications.

Mobile app development has advanced significantly since the original launch of Google Play and the App Store. Nowadays, users expect excellence in design, performance, security, and user experience; it is no longer sufficient just to produce a functional app. This e-book will help you become proficient in the innovative techniques and industry best practices needed to achieve these expectations and enhance your knowledge of developing mobile apps.

You are already aware of the potential and strength of mobile apps, whether you are a seasoned developer or an aspiring app maker. You've mastered the fundamentals of creating mobile apps, giving you a base to grow. With the aid of this e-book, you will be able to go beyond the fundamentals and get the skills and knowledge necessary to produce exceptional apps that stick out in the crowded app store.

In these chapters, you will delve into complex subjects, including advanced UI/UX design principles and intricate architectural patterns. You will explore the complexities of networking, data management, and performance optimization to ensure your apps execute smoothly and are exceptional. Additionally, you'll learn about testing strategies, security precautions, and user interaction strategies that will improve the legitimacy of your app and increase user satisfaction.

You will get the skills and knowledge necessary to overcome obstacles and take advantage of opportunities as we take you on an advanced tour through the world of mobile app development. This e-book's chapters will each reveal a new level of information while giving you helpful advice, real-world examples, and practical ideas you can use right now on your projects.

The concepts and procedures described here will assist you in exceeding expectations and achieving your objectives, whether you are an independent developer looking to make the next big thing or a member of a development team working on enterprise programs that are vital to the organization. After finishing this e-book, you will have the expertise to create mobile applications that adhere to industry requirements and break new ground in terms of excellence.

Together, let's go on this adventure to understand the complexities of developing advanced mobile applications. Get ready to become an expert in creating captivating mobile experiences that increase engagement, enthrall consumers, and leave a lasting impression. Your path to mobile app development mastery starts now.

CHAPTER I

Advanced Mobile App Architecture

Understanding advanced app architecture patterns

Creating robust, scalable, and maintainable applications is paramount in the ever-evolving landscape of mobile app development. To achieve these goals, developers often turn to advanced app architecture patterns. These architectural paradigms offer a structured approach to organizing code and separating concerns, making it easier to manage complexity, ensure code quality, and facilitate collaboration among development teams. In this section, we will delve into the world of advanced app architecture patterns, exploring their significance, common patterns, and how they contribute to the success of modern mobile applications.

One of the fundamental challenges in mobile app development is maintaining a clean, organized, and scalable codebase as the complexity of the application grows. Advanced app architecture patterns solve this challenge by offering well-defined guidelines for structuring an app's code. These patterns are not limited to any specific programming language or platform but can be applied universally. They help developers create applications that are not only functional but also maintainable and adaptable to changing requirements. At the heart of advanced app architecture is the principle of separation of concerns. This principle advocates for breaking down an application into distinct modules or layers, each responsible for a specific aspect of the app's functionality. By separating concerns, developers can make changes or improvements to one part of the application without affecting the others, which enhances code reusability and maintainability.

One of the most widely adopted advanced app architecture patterns is the Model-View-ViewModel (MVVM) pattern. MVVM separates an app's components into three main categories: the Model, the View, and the ViewModel. The app's data and business logic are represented by the Model, the View handles the user interface and presentation, and the ViewModel acts as an intermediary between the Model and the View, translating data and events between them.

MVVM promotes a clean separation of concerns, making it easier to write unit tests, as each component can be tested independently. It also enables better collaboration between designers and developers since the View can be designed separately from the underlying logic. Moreover, MVVM enhances code reusability, as ViewModel instances can often be reused across different parts of an application.

Another popular architecture pattern in mobile app development is the Model-View-Controller (MVC) pattern. While MVC has been around for decades and has proven effective in various contexts, it is still a valuable pattern in modern app development. In MVC, the data and business logic are represented by the Model, the data is shown to the user in the View, and the Controller handles user input and communicates between the Model and View.

MVC has its advantages, such as simplicity and familiarity, as many developers are already acquainted with this pattern. However, it can sometimes lead to massive View Controllers or Activities, which can become hard to manage as an app grows in complexity. To address this limitation, variations of MVC, such as Model- View-Presenter (MVP) and Model-View-Controller- ViewModel (MVCVM), have emerged, providing more structured alternatives.

The patterns mentioned above are just a glimpse into the world of advanced app architecture. Depending on the requirements of a particular project or the preferences of a development team, other patterns like Clean Architecture, Redux, or Flux might be more suitable. These patterns offer different approaches to handling state management, data flow, and the organization of code within an app.

One of the significant benefits of adopting advanced app architecture patterns is the enhancement of code quality. By enforcing a clear separation of concerns and providing guidelines on how data should flow through an application, these patterns reduce the likelihood of bugs and make it easier to detect and fix issues when they arise. Moreover, they facilitate code reviews and collaboration among team members, as everyone follows a standardized structure and naming conventions.

Scalability is another critical aspect where advanced app architecture shines. As mobile applications grow in complexity, scalability becomes a fundamental concern. Architectural patterns like MVVM, MVP, or Clean Architecture offer mechanisms for handling increased complexity without sacrificing performance or code quality. This scalability is crucial, especially for applications needing to accommodate new features or support a growing user base.

Maintainability is closely tied to scalability. A maintainable app architecture allows development teams to make updates, add features, or fix issues efficiently. With a well-structured codebase and clear separation of concerns, developers can work on different parts of the application in parallel, reducing the risk of conflicts and streamlining the development process. This is particularly important for apps with long lifespans that must evolve continuously.

In addition to code quality, scalability, and maintainability, advanced app architecture patterns contribute to the overall productivity of development teams. They provide a common language and structure for developers to work within, making it easier for new team members to understand and contribute to existing codebases. Moreover, these patterns often come with well-established best practices and tools that can streamline development processes, reducing the time and effort required to build and maintain mobile applications.

While advanced app architecture patterns offer numerous benefits, it's important to note that choosing the correct pattern depends on the specific needs of a project. Not all patterns are suitable for every app, and sometimes a combination of patterns may be the best approach. Development teams must carefully assess their project's requirements, constraints, and goals before selecting an architecture pattern.

In conclusion, advanced app architecture patterns are pivotal in modern mobile app development. They provide a structured approach to organizing code, separating concerns, and improving code quality, scalability, and maintainability. By adopting these patterns, developers can create mobile applications that meet functional requirements and excel in user experience and long-term sustainability. As the mobile app development landscape evolves, mastering these architecture patterns becomes increasingly essential for developers aiming to build outstanding and successful mobile applications.

MVVM, MVP, and other architectural models

In the dynamic world of mobile app development, choosing the right architectural model is akin to selecting the blueprint for a skyscraper. It lays the foundation for how an app's code is structured, how data flows, and how the user interface communicates with underlying business logic. Two of the most prominent architectural patterns in this realm are MVVM (Model-View-ViewModel) and MVP (Model-View-Presenter), each with its unique approach to separating concerns and improving code maintainability. However, the architectural landscape is rich and diverse, with various models like Clean Architecture, Redux, and Flux vying for developers' attention. In this section, we will delve into the world of MVVM, MVP, and several other architectural models, exploring their principles, advantages, and considerations when choosing the most suitable one for a particular project.

MVVM, an acronym for Model-View-ViewModel, has gained immense popularity in recent years, primarily due to its suitability for developing robust and maintainable mobile applications. At its core, MVVM emphasizes the separation of concerns, allowing developers to disentangle the user interface (View) from the underlying business logic (ViewModel) while maintaining a clear distinction between data (Model). This separation simplifies the development process by enabling parallel work on the user interface and logic components. The ViewModel acts as an intermediary, facilitating communication between the View and the Model.

One of the key advantages of MVVM is its testability. By isolating the business logic in the ViewModel, developers can write unit tests more effectively, verifying that the logic functions correctly without requiring the presence of the user interface. This promotes a more rigorous approach to quality assurance and reduces the likelihood of introducing bugs during development.

Moreover, MVVM supports data binding, a powerful feature that automates data synchronization between the Model and the View. When data in the ViewModel changes, it automatically updates the associated View, and vice versa. This seamless data flow enhances the responsiveness of the user interface and simplifies the code responsible for updating UI components.

Despite its benefits, developers unfamiliar with the pattern may find MVVM to have a more difficult learning curve, particularly when working with complex data-binding setups. Furthermore, implementing MVVM effectively may require a comprehensive understanding of the underlying frameworks and libraries, which can pose challenges for those not well-versed in the chosen technology stack.

While MVVM represents a more recent addition to the app architecture landscape, Model-View-Presenter (MVP) is a classic and reliable architectural pattern that has withstood the test of time. In MVP, the Model represents the data and business logic, the View manages the user interface and user input, and the Presenter acts as an intermediary, handling communication between the Model and the View. The key distinction from MVVM is that the Presenter takes a more active role in managing the View, often explicitly controlling the UI updates.

One of the chief advantages of MVP is its simplicity and ease of understanding. Developers transitioning from traditional desktop application development often find MVP's structure familiar and intuitive. This makes it a desirable option for projects involving a large team of developers with a range of skill levels or with limited time for development.

Additionally, MVP can provide fine-grained control over the user interface, which can be advantageous in scenarios requiring precise control over UI updates. This control can help optimize performance by preventing unnecessary updates to UI components.

However, MVP does have its limitations. The increased coupling between the View and Presenter can lead to more complex code, making it challenging to test the View in isolation. Additionally, the increased interaction between the Presenter and the View can sometimes result in larger and more convoluted Presenter classes, making maintenance and code readability a concern as an app grows in complexity.

Moving beyond MVVM and MVP, Clean Architecture is a design philosophy that promotes the creation of applications with well-defined layers, each responsible for specific concerns. This approach goes beyond the immediate separation of Model, View, and Presenter/ViewModel and introduces additional layers such as the Entity, Use Case, and Repository layers.

The primary goal of Clean Architecture is to maximize code maintainability and independence from external frameworks and libraries. It allows for developing apps that can be easily adapted to changing requirements or ported to different platforms without major code overhauls. By relying on interfaces and abstractions,

Clean Architecture enables the replacement of components with minimal impact on the rest of the system.

One of the crucial advantages of Clean Architecture is its long-term sustainability. Applications built following this approach tend to be more resistant to the erosion of technical debt and can evolve over time without architectural bottlenecks. This is particularly valuable for apps with extended lifecycles or those that need to adapt to new technologies and platforms.

However, Clean Architecture may introduce an initial overhead in terms of development time and complexity, especially for smaller projects or tight deadlines. Additionally, understanding and implementing Clean Architecture correctly may require a solid grasp of design principles and software engineering practices.

Architectural patterns like Redux and Flux shine when dealing with state management in complex applications. These patterns focus on unidirectional data flow, ensuring that changes to the application state are predictable and well-managed. They are prevalent in web and mobile applications that rely heavily on dynamic user interfaces and real-time updates.

In Redux and Flux, the state of an application is stored in a central repository, often referred to as a store. Components can dispatch actions to modify the state, and the changes are propagated through a series of reducers. This ensures that state changes are predictable and traceable, making debugging and testing easier.
One of the key advantages of Redux and Flux is their ability to manage complex state transformations gracefully. This makes them well-suited for applications

that require real-time updates, such as social media platforms, e-commerce sites, and collaborative tools. The strict unidirectional data flow simplifies the tracking of data changes and helps prevent unpredictable side effects.

However, implementing Redux or Flux can introduce additional complexity, especially in smaller applications where the benefits of strict state management might not outweigh the development overhead. Furthermore, developers who are not familiar with the ideas behind stores, actions, and reducers may find the learning curve to be quite steep.

Selecting the most appropriate architectural model for a mobile app project is a critical decision that can significantly impact the development process and the long-term maintainability of the application. The choice often hinges on several factors, including the project's complexity, the development team's familiarity with the chosen model, and the application's specific requirements.

MVVM and MVP offer potent solutions for separating concerns in mobile app development, each with its strengths and considerations. MVVM excels in complex applications that require extensive data binding and a clear separation of concerns. Conversely, MVP is a reliable choice for projects with simpler requirements or when developers favor a more traditional approach.

Clean Architecture is an excellent choice for projects with a focus on long-term sustainability, independence from external dependencies, and adaptability to changing requirements. It's particularly valuable for large-scale applications or those with extended lifecycles.

Redux and Flux shine in managing complex application states, making them indispensable for applications that rely heavily on real-time updates and dynamic user interfaces.

In conclusion, architectural models like MVVM, MVP, Clean Architecture, Redux, and Flux provide a structured approach to app development, each with its unique set of advantages and considerations. The right model choice should align with the project's goals, team expertise, and specific technical requirements. Regardless of the chosen architectural model, a well-thought-out and well-executed architecture is crucial for building mobile apps that are not only functional but also scalable, maintainable, and adaptable to the ever-evolving world of mobile technology.

Building scalable and maintainable apps

Creating an app is just the beginning of a much longer journey in the rapidly evolving landscape of mobile app development. To ensure an app's long-term success, it must be scalable and maintainable. Scalability ensures that the app can grow to meet increasing user demands, while maintainability ensures that it can evolve with changing requirements and technology landscapes. In this section, we will explore the significance of building scalable and maintainable apps, the challenges developers face in achieving these goals, and the strategies and best practices that can pave the way for sustainable app development.

Scalability and maintainability are two essential pillars of successful app development. Scalability ensures that an app can handle a growing user base, increased data volume, and new features without compromising performance or user experience. When an app gains popularity and attracts a larger audience, it must be able to scale seamlessly to meet the demand, or risk losing users due to slow load times, crashes, or other performance issues.

Maintainability, conversely, ensures that an app remains adaptable and manageable in the face of evolving requirements, technology updates, and bug fixes. A maintainable app can easily accommodate new features, integrate with emerging technologies, and undergo routine maintenance without causing significant disruptions to the user experience. Without proper maintainability, apps can become outdated, prone to security vulnerabilities, and expensive to manage and update.

Building a scalable app has its challenges. One of the primary hurdles is predicting future growth accurately. Overestimating or underestimating scalability needs can lead to inefficient resource allocation and either overspending on infrastructure or facing performance bottlenecks. Balancing the cost of scalability with the expected user growth is a delicate task.

Another challenge is designing an architecture that can handle increased loads. Traditional monolithic architectures, where the entire app is bundled, may struggle to scale efficiently. On the other hand, Microservices architecture divides the app into smaller, independently deployable components, making it easier to scale only the parts of the app that need it.

Additionally, managing data growth can be challenging. Scalable apps must handle large datasets, and developers must consider how to store and retrieve data efficiently, especially as the volume of data increases over time.

Strategies like database sharding and caching can help address these challenges.

Several strategies can aid in building scalable apps. One of the fundamental principles is designing for horizontal scalability. This means that instead of scaling up by adding more resources to a single server (vertical scaling), the app can handle increased load by adding more servers (horizontal scaling). Cloud services like AWS, Azure, and Google Cloud offer auto-scaling features that automatically adjust resources based on demand. Another approach is implementing a content delivery network (CDN) to distribute content geographically. CDNs can cache and serve static assets like images and videos from servers located closer to the end-users, reducing latency and speeding up content delivery.

Furthermore, asynchronous processing can help offload resource-intensive tasks from the main application thread. Non-essential operations, such as processing image uploads, sending emails, and creating reports, can be completed without degrading the responsiveness of the service by using message queues and background workers.

Maintainability poses its own set of challenges. One common issue is code complexity. As an app evolves and new features are added, the codebase can become convoluted and difficult to navigate. This complexity can slow down development, make it harder to find and fix bugs and increase the risk of introducing new issues when making changes.

Another challenge is managing technical debt. Technical debt accumulates when shortcuts or less-than-optimal solutions are used to meet deadlines or address immediate needs. Over time, this debt can slow down development, increase the likelihood of bugs, and make implementing new features or technologies harder.

Additionally, maintaining consistent documentation is crucial. As developers come and go, or as the app evolves, documentation can serve as a critical resource for understanding how the app works, its architecture, and best practices for development. Lack of documentation can hinder onboarding new team members and make maintenance tasks more challenging.

Maintaining an app's health and adaptability requires several strategies and best practices. Code refactoring, the process of restructuring existing code without changing its external behavior, is essential for reducing complexity and addressing technical debt. Regularly reviewing and refactoring code can improve its readability, maintainability, and extensibility.

Implementing a robust version control system (VCS), such as Git, is also crucial. Developers can work together efficiently, keep track of changes made to the codebase, and undo changes when problems occur due to VCS. Additionally, it lowers the possibility of introducing regressions by allowing developers to work independently on new features or bug solutions.

Pipelines for continuous integration and continuous delivery, called CI/CD, speed up the testing, building, and deploying of code changes. With automated testing and deployment, developers can catch issues early and ensure that updates are delivered to users quickly and reliably.

In terms of documentation, maintaining up-to-date and comprehensive documentation is essential. This includes documenting code, APIs, database schemas, and architectural decisions. Proper documentation ensures that developers have the necessary information to work on the app and that new team members can quickly become productive contributors.

Building scalable and maintainable apps is not just a development task; it's a long-term investment in an app's success. Scalability ensures that an app can handle growth and increased demand, while maintainability ensures that it can evolve, adapt, and remain reliable over time. While challenges abound in achieving these goals, the rewards are substantial. Strategies like horizontal scalability, asynchronous processing, and content delivery networks can address scalability challenges, while code refactoring, version control, CI/CD pipelines, and comprehensive documentation are essential for maintaining apps effectively.

Scalable and maintainable apps are not a luxury; they are a need in the digital world where user expectations are constantly rising and technological advancements are happening quickly. Developers and organizations prioritizing these qualities are better equipped to thrive in the competitive world of mobile app development, deliver exceptional user experiences, and achieve long-term success.

Case studies of successful apps with advanced architectures

In mobile app development, advanced architectural patterns have become instrumental in shaping the success and sustainability of modern applications. To appreciate the true impact of these architectures, it's worth examining some real-world case studies where advanced architectural choices played a pivotal role in creating exceptional and widely-admired mobile apps.

A social media giant, Instagram is an excellent example of how advanced architecture can drive success. Instagram's app architecture has evolved over the years, but a key turning point was their migration to a combination of Python, Django, and PostgreSQL on the server-side, and a lean native app on the client side.

One of the standout architectural features is Instagram's use of a distributed database system that allows for seamless scaling to accommodate millions of users. This architectural choice allowed the app to maintain its fast performance and responsiveness, even as it grew exponentially in user numbers and content uploads.

Furthermore, Instagram leverages a streamlined and minimalistic user interface, aligning perfectly with the Model-View-Controller (MVC) architectural pattern. This approach ensures the user experience remains consistent, intuitive, and visually appealing. By embracing both scalability and simplicity, Instagram attracted users and retained them, ultimately leading to its global success.

The ride-sharing giant, Uber relies heavily on a sophisticated app architecture that integrates cutting-edge geolocation and real-time optimization technologies. A microservices-based approach at the core of Uber's architecture allows for agility and scalability.

Uber's system utilizes geolocation services to pinpoint users' locations, track drivers in real time, and optimize routes. The architecture employs a combination of technologies, including Node.js, Python, and Go, to rapidly process large volumes of location data and provide users with real-time ride information.

Additionally, Uber's app is designed with a focus on responsiveness, powered by an Event-Driven Architecture (EDA) that ensures prompt handling of user requests and dynamic updates of ride availability and prices. This architectural choice enables Uber to deliver an exceptional and efficient experience to users worldwide, making it one of the most successful ride-sharing platforms globally.

Airbnb, the online marketplace for lodging and travel experiences, showcases the power of microservices architecture in delivering a seamless and user-friendly experience. Airbnb's app relies on a complex ecosystem of microservices, each responsible for a particular function, such as booking management, user authentication, and recommendation algorithms.

By adopting microservices, Airbnb can rapidly innovate and iterate on individual components of its app without affecting the entire system. This flexibility is crucial for a platform that must accommodate various accommodations, host profiles, and constantly changing market dynamics.

Moreover, Airbnb's app leverages advanced algorithms for search and recommendations, delivering personalized travel experiences to users. These algorithms are supported by a data-driven architecture that collects and processes vast amounts of user data, ensuring that the platform provides relevant and enticing suggestions to travelers.

In conclusion, these case studies highlight the pivotal role that advanced app architectures play in the success of modern mobile applications. Instagram's scalability and simplicity, Uber's real-time optimization and geolocation capabilities, and Airbnb's microservices-driven flexibility and personalized recommendations all demonstrate the potential of advanced architectures to elevate user experiences, drive user engagement, and ultimately lead to the widespread adoption and success of mobile apps. Selecting the right architectural model and technology might be crucial to attaining sustainable growth and preserving a competitive edge in today's competitive app industry.

CHAPTER II

Advanced UI/UX Design Principles

Designing for a great user experience

In mobile app development, one factor reigns supreme: the user experience (UX). Creating a mobile app that functions flawlessly and delights and engages users is essential for success. Whether you're developing a utility app, a game, or a productivity tool, the design of your app is a critical determinant of its popularity and longevity. In this section, we will explore the basic principles and strategies for designing a great user experience in mobile app development and how these principles can be applied to create user-centric, intuitive, and visually appealing apps.

User-centered design is the foundation of a great UX. It begins with understanding your target audience's needs, preferences, and pain points. Conducting user research, surveys, and user testing can provide invaluable insights into what your users expect from your app. By empathizing with your users, you can design an app that addresses their problems and offers a tailored solution.

Navigation is a crucial aspect of UX. Users should be able to navigate your app effortlessly, find what they need quickly, and complete tasks without confusion. Keep navigation menus clean and well-organized, use clear and familiar icons, and minimize the steps required to perform common actions. Intuitive navigation enhances user satisfaction and encourages prolonged engagement with your app.

Mobile devices come in various sizes and orientations. A responsive design ensures your app adapts seamlessly to different screen sizes and resolutions. Whether it's a smartphone, tablet, or even a wearable device, users expect consistent and visually appealing experiences across all platforms. Responsive design improves usability and demonstrates a commitment to user convenience and accessibility.

App speed is a fundamental component of user experience. Slow-loading screens, laggy interactions, or unresponsive buttons can frustrate users and drive them away. To ensure a snappy and responsive app, optimize images and assets, use lazy loading for non-essential content, and employ efficient algorithms and data structures. Performance optimization enhances user satisfaction and encourages longer app usage.

Consistency in branding and design elements is essential for building a memorable and trustworthy app. Maintain a cohesive color scheme, typography, and visual style throughout the app. Your app's branding should align with your company's or product's identity, reinforcing recognition and trust. Consistent branding fosters a sense of familiarity and reliability among users.

Accessibility is not an afterthought; it should be an integral part of your app design process. Ensure your app is accessible to users with disabilities, such as those with hearing, visual, or motor impairments. Implement features like alt text for images, keyboard navigation, and voice commands. Having an accessible app shows that you value inclusion and social responsibility while also growing your user base.

The design process continues after the app's launch. Collect user feedback through reviews, surveys, and analytics. Pay attention to user suggestions and pain points, and be prepared to iterate and improve your app based on this feedback. Continuous improvement is vital for staying competitive and maintaining user engagement over time.

Usability testing involves having real users interact with your app and providing feedback on its usability. It's a critical step in ensuring that your app aligns with user expectations and resolves usability issues before they become widespread. Usability testing aids in identifying areas for improvement, refining your design, and enhancing the user experience as a whole.

In the fiercely competitive world of mobile app development, designing for a great user experience is not just a goal; it's a competitive advantage. User-centric design, intuitive navigation, responsive layouts, performance optimization, consistent branding, accessibility, continual feedback, and usability testing are all essential components of creating an app that satisfies user needs and exceeds their expectations.

A great user experience is not just a one-time achievement but an ongoing commitment to delivering value and satisfaction to your users. By prioritizing user experience in your mobile app development process, you can create apps that resonate with users, drive engagement, and stand the test of time in an ever-evolving digital landscape.

Advanced UI design patterns

User Interface (UI) design is critical to mobile app development. The way an app looks and feels dramatically influences user satisfaction and engagement. Developers and designers often turn to advanced UI design patterns to create genuinely exceptional mobile apps. These patterns go beyond the basics of buttons and layouts, offering innovative and user-centric solutions for crafting compelling and intuitive interfaces. This section will explore several advanced UI design patterns and their significance in mobile app development.

The card-based design has gained immense popularity, thanks partly to platforms like Pinterest and Google's Material Design. This approach involves presenting information in self-contained "cards," each typically representing a single piece of content. The card-based design offers several advantages. First, it provides a clear visual hierarchy, making it easy for users to scan and prioritize content. Second, it promotes information chunking, breaking down complex content into digestible units. Finally, it adapts well to mobile devices, where screen real estate is limited, allowing users to swipe or scroll through cards effortlessly.

Gestures and microinteractions are subtle yet powerful UI design patterns that enhance user engagement. Gestures, such as swiping, pinching, or tapping, enable users to interact with an app more naturally and intuitively. They provide shortcuts for navigation, content interaction, and actions. Microinteractions, on the other hand, are small, delightful animations or feedback mechanisms that respond to user actions. For example, a subtle animation when liking a post or a satisfying "ding" when sending a message. These microinteractions provide feedback, create a sense of responsiveness, and make the app experience more enjoyable.

A design pattern called progressive disclosure helps manage complex interfaces by gradually releasing information in response to user requests or actions. Rather than overwhelming users with a barrage of options or details upfront, the app presents only the essential information, allowing users to explore further if desired. For example, an email app might initially display a list of subject lines and sender names, revealing the entire email content when a user taps on a specific message. This pattern streamlines the user experience, reduces cognitive load, and keeps users engaged without inundating them with unnecessary information.

Bottom navigation is a mobile-specific UI pattern that places navigation elements at the bottom of the screen, within easy reach of the user's thumb. This pattern is beneficial in larger smartphones, where one-handed use is common. Bottom navigation simplifies access to primary app sections, such as home, search, and profile, without requiring users to stretch or adjust their grip. It's an excellent example of how UI design can be tailored to the ergonomic realities of mobile devices, enhancing user comfort and convenience.

Dark mode has emerged as a popular UI design trend, offering both aesthetic appeal and functional benefits. This design pattern switches the app's color scheme to a darker background with lighter text and elements. Dark mode reduces eye strain in low-light conditions, conserves battery life on OLED screens, and adds a touch of modern elegance to an app's appearance. It's not just a trend; it's a user-driven design choice that caters to diverse user preferences and needs.

The Floating Action Button (FAB) is a UI element that typically appears as a circular button with an iconic symbol. It floats above the app's content and serves as a focused call to action, often associated with the app's primary task or a frequently performed action. For example, in a note-taking app, the FAB could be used for adding new notes. The FAB is a prime example of a design pattern that encourages users to take specific actions, making it easier to complete key tasks within the app.

Split-screen layouts are a valuable UI design pattern in tablet or large-screen mobile devices. This approach divides the screen into two or more distinct areas, allowing users to interact with multiple app sections or functions simultaneously. For example, a split-screen layout in a productivity app might enable users to view a document on one side while taking notes on the other. This pattern supports multitasking and enhances user efficiency, making it particularly relevant in contexts where productivity is a priority.

In conclusion, advanced UI design patterns in mobile app development are more than just visual aesthetics; they are tools for improving usability, engagement, and overall user satisfaction. By embracing these patterns, app developers and designers can create interfaces that look modern, function intuitively, and delight users. In an increasingly competitive mobile app landscape, these advanced UI design patterns play a vital role in differentiating an app, retaining users, and ultimately contributing to its success.

Responsive design and adaptive layouts

Mobile app development has evolved significantly, driven by the diversity of devices and screen sizes. Developers turn to responsive design and adaptive layouts to create apps that provide a consistent and user-friendly experience across this multitude of devices. These design approaches are essential for ensuring that mobile apps function correctly and look great on a wide range of screens. This section will delve into responsive design and adaptive layouts, exploring their significance, key differences, and applications in mobile app development.

Responsive design focuses on creating web and mobile app layouts that adapt fluidly to different screen sizes and orientations. Offering the best possible viewing experience on all devices—smartphone, tablet, laptop, or desktop computer—is the main objective of responsive design. Responsive design achieves this by combining flexible grids, media queries, and scalable images and fonts.

The use of percentage-based layouts rather than fixed pixel values is one of the basic features of responsive design. Elements such as images, text, and containers are sized proportionally to the screen's width, ensuring they adjust automatically to fit the available space. Media queries are CSS techniques that allow developers to define particular styles for different screen sizes or breakpoints. By using media queries, designers can reorganize content, change font sizes, or hide some aspects as the screen size changes.

For example, a responsive website may display a single-column layout with large text and buttons on a mobile device, while the same site on a desktop screen might feature a multi-column layout with smaller fonts and additional content visible simultaneously.

On the other hand, adaptive layouts take a more deliberate approach to catering to different device categories. Instead of creating a single, flexible design that adjusts to all screens, adaptive layouts involve creating distinct layouts tailored for specific device categories or screen sizes. These layouts are designed to deliver an optimized experience for each type, such as smartphones, tablets, or desktops.

Adaptive layouts typically involve designing and developing multiple versions of an app's user interface, each optimized for a particular screen size or type of device. For instance, an e-commerce app may have separate layouts for smartphones, focusing on streamlined product browsing and easy checkout, and tablets, which can use larger screens to display more products and additional navigation options.

The primary difference between responsive design and adaptive layouts is their handling of screen diversity. Responsive design creates a single, flexible layout that adapts to different screens using fluid grids and media queries. In contrast, adaptive layouts involve creating distinct, tailored layouts for specific device categories.

Several factors come into play when deciding between responsive design and adaptive layouts. The choice depends on the app's complexity, the target audience, development resources, and the desired level of customization. Responsive design is often more straightforward to implement and is suitable for content-driven apps or when a consistent design aesthetic is essential across all devices. Adaptive layouts are a better fit for apps that require a high degree of optimization and customization for various device categories.

In mobile app development, responsive design and adaptive layouts are potent tools for delivering a unified user experience across the diverse landscape of mobile and web devices. While responsive design offers flexibility and adaptability through fluid grids and media queries, adaptive layouts take a more targeted approach by tailoring interfaces to specific device categories.

Ultimately, the choice between responsive design and adaptive layouts depends on the project's goals, complexity, and budget. Regardless of the strategy used, the primary objective is always to provide users a smooth and visually appealing experience, no matter the device they use to access the app. By mastering these design techniques, mobile app developers can ensure their creations are accessible and enjoyable to users on a wide range of screens, contributing to the app's success and user satisfaction.

CHAPTER III

Advanced Development Tools and IDEs

Exploring advanced development environments

Mobile app development is a dynamic field that continually evolves to meet the demands of a tech-savvy world. Advanced mobile app development environments have emerged as essential tools for developers, enabling them to create powerful, feature-rich, cross-platform applications efficiently. In this section, we will explore some advanced development environments, discussing their significance and how they streamline the app development process.

The open-source UI software development toolkit Flutter, created by Google, has become quite popular because of its capability of producing natively compiled desktop, web, and mobile applications from a single codebase. At the heart of Flutter's appeal are its widgets, which are building blocks for constructing user interfaces. Flutter's widget-centric architecture allows developers to create visually stunning and customizable apps by composing widgets. The "hot reload" feature provides real-time feedback, enabling developers to see changes instantly, facilitating rapid iteration and experimentation. Flutter's commitment to delivering a native experience on multiple platforms has made it a go-to choice for building cross-platform apps with exceptional performance.

React Native, maintained by Facebook, is another prominent player in the cross-platform mobile app development landscape. It grants developers to develop mobile apps using JavaScript and React, a popular web development library. React Native employs a "write once, run anywhere" philosophy, enabling developers to reuse a substantial portion of their codebase across iOS and Android platforms. This approach accelerates development and reduces maintenance overhead. React Native's extensive library of pre-built components and third-party plugins further streamlines the development process. Many well-known apps, including Facebook, Instagram, and Airbnb, have been built using React Native, attesting to its capability to deliver native-like performance and user experiences.

Xamarin, acquired by Microsoft, offers a unique approach to cross-platform mobile app development by using C# and the .NET framework. Xamarin allows developers to share code, business logic, and user interface elements across iOS and Android platforms while delivering native performance. Xamarin.Forms, a UI toolkit included in Xamarin, simplifies the creation of a single user interface for both platforms, reducing development time and ensuring a consistent look and feel. The tight integration with Visual Studio, Microsoft's integrated development environment, provides developers with a comprehensive suite of tools for efficiently building, testing, and debugging apps.

Kotlin Multiplatform is an innovative approach to cross-platform app development that leverages Kotlin, a modern and expressive programming language. Developed by JetBrains, Kotlin Multiplatform allows developers to share code between Android and iOS apps while maintaining a high degree of platform-specific optimization. Unlike some other cross-platform solutions, Kotlin Multiplatform aims to provide a more flexible balance between code sharing and platform-specific customization. This approach allows developers to create highly performant apps that remain closely aligned with the native development experience for each platform.

Using Angular, TypeScript, and JavaScript, developers can build native mobile apps with the help of the open-source NativeScript framework. It provides direct access to native APIs, allowing for a truly native user experience while still using web technologies. NativeScript leverages a single codebase for iOS and Android, making it an attractive choice for teams with web development expertise. The framework offers extensive support for third-party libraries and plugins, further extending its capabilities and simplifying complex tasks.

Advanced mobile app development environments have revolutionized how developers create cross-platform apps, making the process more efficient and accessible. Flutter, React Native, Xamarin, Kotlin Multiplatform, and NativeScript each offer unique advantages, catering to various development preferences and needs. These environments empower developers to bring their app ideas to life, regardless of their expertise in native languages or platforms. As mobile app development continues to evolve, these advanced environments will play a vital role in shaping the future of app development, enabling developers to deliver high-quality, feature-rich, and cross-platform apps easily.

Best practices for using integrated development environments

Integrated Development Environments (IDEs) are indispensable tools for mobile app developers, providing a centralized platform for coding, debugging, testing, and deploying applications. Developers should adhere to best practices when using IDEs to maximize productivity and create high-quality mobile apps. This section will explore these best practices and how they contribute to efficient and effective mobile app development.

The first step in leveraging an IDE effectively is selecting the right one for your needs. Different IDEs cater to various mobile app development platforms, such as Android Studio for Android, Xcode for iOS, Visual Studio for Xamarin, and Visual Studio Code for cross-platform development. Assess your project requirements, target platforms, and programming languages to choose an IDE that aligns with your goals.

IDEs frequently receive updates introducing bug fixes, performance enhancements, and new features. Staying up-to-date with the latest version of your chosen IDE is crucial to benefit from these improvements. Regular updates ensure compatibility with the latest mobile operating system versions and development tools.

IDEs offer a range of customization options to tailor the development environment to your preferences. Customize keyboard shortcuts, code formatting settings, color themes, and code templates to create a workspace that suits your workflow. A personalized workspace can significantly enhance productivity by reducing unnecessary friction during development.

Efficient code navigation and search features are essential for large-scale mobile app projects. IDEs provide tools to quickly find and navigate to specific classes, functions, or variables within your codebase. Familiarize yourself with these features to save time and maintain code readability.

Integrating version control systems like Git directly into your IDE streamlines collaborative development and code management. IDEs often offer features like commit, pull, and push directly from the interface, making it easier to track changes, resolve conflicts, and maintain code quality through version control.

Effective IDEs include code refactoring tools that help developers improve code quality and maintainability. Utilize features like renaming variables, extracting methods, and optimizing imports to enhance your codebase. Additionally, leverage code analysis and inspection tools to identify and rectify potential issues and maintain code consistency.

Debugging is an integral part of mobile app development. IDEs provide powerful debugging tools that allow you to set breakpoints, inspect variables, step through code execution, and identify and fix bugs efficiently. Familiarize yourself with these tools to expedite the debugging process.

Automated testing is essential for ensuring app functionality and stability. IDEs often integrate with testing frameworks and offer features for running unit, UI, and performance tests directly from the development environment. Implementing and regularly running tests within your IDE can capture issues early in the development cycle.

IDEs often support plugins and extensions that extend their functionality. Explore available plugins and extensions to enhance your development environment. These can include additional code generators, linters, project management tools, and integrations with third-party services.

Mastering keyboard shortcuts in your IDE can significantly boost productivity. Learn the essential shortcuts for code navigation, refactoring, debugging, and other everyday tasks. Many IDEs offer built-in cheat sheets or allow you to customize shortcuts to match your workflow.

IDEs often come with extensive documentation and learning resources. Take advantage of tutorials, documentation, forums, and online communities to effectively expand your knowledge and troubleshoot issues.

In mobile app development, using an Integrated Development Environment can make a substantial difference in project efficiency and code quality. By adhering to best practices like selecting the right IDE, staying updated, customizing your workspace, and leveraging essential features, developers can streamline their workflows, reduce development time, and create mobile apps that meet high functionality and user experience standards. IDEs are indispensable allies for mobile app developers, empowering them to bring their app ideas to life efficiently and effectively.

Leveraging code version control systems

Code version control systems (VCS) play a pivotal role in modern software development, and mobile app development is no exception. These systems, such as Git,

SVN (Subversion), and Mercurial, provide developers with the means to track changes to their code, collaborate effectively, and ensure the integrity of their projects. In this section, we will explore the significance of leveraging code version control systems in mobile app development and their numerous advantages to development teams.

In mobile app development, multiple developers often work on the same project simultaneously. Code version control systems allow developers to track changes made to the codebase, providing a comprehensive history of who made what modifications, when, and why. This history is invaluable for collaboration, as it enables developers to understand the evolution of the codebase, identify the source of issues or bugs, and coordinate their efforts effectively.

Effective collaboration is a cornerstone of successful mobile app development. Code version control systems facilitate collaboration by providing mechanisms for multiple developers to work on the same codebase concurrently. Features like branching and merging allow developers to create separate branches of the code for specific tasks or features, preventing conflicts and enabling smooth integration of changes into the main codebase.

Code reviews are essential for maintaining code quality and catching potential issues early in development. Code version control systems enable teams to conduct code reviews efficiently by providing a structured platform for reviewing changes, leaving comments, and ensuring that code follows to coding standards and best practices. This process improves code quality and facilitates knowledge sharing and mentoring among team members.

Mobile app development projects represent a substantial investment of time and effort. Code version control systems act as a safeguard by keeping a secure copy of the entire codebase, both locally and on remote servers. This redundancy ensures that code assets are protected against accidental deletions, hardware failures, or data loss, providing a crucial layer of disaster recovery and peace of mind.

Continuous integration (or CI) and continuous delivery (or CD) are integral to modern mobile app development. Code version control systems seamlessly integrate with CI/CD pipelines, allowing for automated testing and deployment. When developers push changes to the VCS, CI/CD systems can trigger automated build processes, execute tests, and deploy the app to testing or production environments, streamlining the development pipeline and ensuring that changes are validated and delivered consistently.

Code version control systems encourage innovation by enabling developers to experiment and branch off the main codebase without disrupting the stability of the project. Developers can create feature branches to work on new functionalities, conduct experiments, or explore alternative solutions. This flexibility fosters creativity and enables teams to pursue innovative ideas without risking the integrity of the core codebase.

Managing releases and versions of a mobile app is simplified with code version control systems. Developers can tag specific commits or versions, making it easy to identify and revert to a particular state of the codebase if issues arise. This ability to rollback to a known working state is invaluable for maintaining a stable app and responding quickly to critical bugs or unforeseen matters in production.

Code version control systems are indispensable tools for mobile app development teams. They empower developers to collaborate seamlessly, maintain code quality, safeguard code assets, streamline development pipelines, and encourage innovation. In a rapidly evolving and collaborative field like mobile app development, leveraging code version control systems is not just a best practice but a cornerstone of efficient, productive, and successful development processes. Developers who embrace these systems are better equipped to deliver high-quality, feature-rich mobile apps that meet user expectations and remain competitive in the dynamic mobile app market.

Advanced debugging and profiling techniques

Debugging and profiling are essential activities in mobile app development that help identify and resolve issues, optimize performance, and enhance the overall quality of applications. While basic debugging and profiling tools are readily available, advanced techniques and tools have emerged to tackle the increasingly complex challenges developers face in the mobile app development landscape. This section will explore these advanced debugging and profiling techniques and their significance in building high-quality, performant mobile apps.

Automated testing and Test-Driven Development (TDD) are foundational techniques contributing to advanced debugging and profiling. Developers can identify and address issues early by writing test cases for individual app components or features before writing the actual code. Automated testing frameworks, such as JUnit for Android and XCTest for iOS, allow developers to automate the execution of test cases, ensuring that the app behaves as expected throughout development. This approach not only catches bugs early but also provides a safety net for future code changes, reducing the risk of regressions.

Remote debugging tools enable developers to inspect and debug mobile apps running on physical devices or emulators remotely. For example, Chrome DevTools Remote Debugging allows developers to debug web-based mobile apps directly from the Chrome browser. Real-time inspection of network traffic, device logs, and application state provides valuable insights into app behavior and performance, even when deployed on users' devices.

Memory-related issues, such as memory leaks and excessive memory usage, can severely impact app performance and stability. Advanced memory profiling tools, like Android's Memory Profiler and Xcode's Instruments for iOS, allow developers to track memory allocations, detect memory leaks, and optimize memory usage. These tools provide detailed memory usage reports, making identifying and resolving memory-related issues easier.

Network-related issues can significantly impact the user experience, especially in mobile apps that rely on data synchronization. Network profiling tools like Charles Proxy and Wireshark enable developers to inspect network requests, responses, and latencies. Advanced tools also provide performance analysis features, helping developers identify bottlenecks, optimize API calls, and reduce data transfer overhead.

CPU profiling tools help developers analyze the execution time of various code segments within their apps. Profiling tools, such as Android Profiler and Instruments for iOS, allow developers to identify CPU-intensive functions and optimize them for better performance. By profiling CPU usage, developers can pinpoint performance bottlenecks and ensure the app runs smoothly, even on devices with limited processing power.

Advanced User Interface (UI) inspection tools, like the Accessibility Inspector on iOS and UI Automator Viewer on Android, enable developers to examine the app's user interface hierarchy and accessibility attributes. This inspection helps ensure that the app is accessible to users with disabilities and that UI elements are correctly rendered and responsive. UI testing frameworks, such as Appium and Espresso, facilitate automated UI testing to verify that the app's interface functions as expected across different devices and screen sizes.

Integrating crash reporting and analytics tools, such as Crashlytics (now part of Firebase) and Flurry, into mobile apps, provides valuable data on app crashes, exceptions, and user interactions. These tools offer advanced features like crash stack traces, user segmentation, and real-time monitoring. By leveraging these insights, developers can quickly identify the root causes of crashes and prioritize bug fixes based on their impact on users.

Advanced debugging and profiling techniques are essential for mobile app developers striving to deliver high-quality, performant applications. These techniques enable developers to catch and resolve issues early in development, optimize app performance, and enhance the overall user experience. In an ever-evolving mobile app landscape, where user expectations are continually rising, leveraging these advanced debugging and profiling tools and practices is essential for building apps that stand out, perform well, and meet user demands.

CHAPTER IV

Advanced Data Management

Managing data in complex mobile applications

In mobile app development, data is the lifeblood that fuels user experiences and functionality. Managing data effectively is crucial, especially in complex mobile applications where information flows seamlessly between the user and the backend systems. From user profiles and preferences to real-time updates and vast databases, data management challenges in complex mobile apps are multifaceted. This section will explore the intricacies of managing data in such applications, the key considerations, and best practices developers employ to create robust and responsive mobile experiences.

Effective data management starts with thoughtful data architecture and design. This involves defining the data schema, relationships, and storage mechanisms in complex mobile applications. Using databases like SQLite or NoSQL solutions like Firebase or MongoDB, developers structure data to facilitate efficient retrieval and manipulation. Well-designed data architectures optimize data access, reduce redundancy, and ensure data integrity, which are vital for complex apps.

Complex mobile apps often need to function seamlessly even when users are offline or experiencing a weak network connection. Implementing caching mechanisms allows apps to store and retrieve frequently accessed data locally, reducing the need for continuous network requests. Client-side caching and background synchronization are two strategies that provide offline user interaction and access to essential data.

Real-time data synchronization is fundamental for many complex mobile applications, such as messaging apps, collaborative tools, and social networks. Implementing data synchronization mechanisms involves handling conflicts, managing updates, and ensuring data remains consistent across devices and platforms. Solutions like WebSocket and Firebase Realtime Database enable real-time data synchronization, facilitating instant updates and collaboration.

In complex mobile applications, data security is paramount. Sensitive user information, financial data, and confidential content require robust encryption and authentication mechanisms to safeguard against unauthorized access and data breaches. Employing encryption protocols like SSL/TLS and implementing user authentication with OAuth or OpenID Connect are essential to protecting user data.

The need for scalable data management becomes apparent as mobile applications grow in complexity and user base. Developers must design systems that can handle increasing data volumes and concurrent user requests without sacrificing performance. Utilizing scalable cloud-based databases, content delivery networks (CDNs), and load balancing techniques ensures that complex apps can scale as needed to meet user demand.

Data privacy regulations, like GDPR and CCPA, impose strict requirements for collecting, storing, and processing user data. Developers of complex mobile apps must adhere to these regulations and implement features like data consent forms, data export options, and user data erasure to maintain compliance. Compliance with data privacy laws avoids legal repercussions and builds user trust.

Developers rely on analytics and user insights to continuously improve complex mobile applications. Implementing analytics tools like Google Analytics or Mixpanel allows developers to track user behavior, monitor app performance, and gather valuable data for decision-making. Analyzing this data helps identify areas for improvement, user engagement trends, and feature adoption rates.

In complex mobile applications, errors and issues are inevitable. Implementing robust error handling mechanisms and monitoring tools like Crashlytics and Sentry helps developers detect, diagnose, and resolve issues promptly. Real-time error tracking and performance monitoring enable developers to provide a seamless user experience and maintain app reliability.

In complex mobile applications, effective data management is not a mere technical task but a fundamental element of user satisfaction, functionality, and business success. It involves careful planning, meticulous design, and continuous monitoring and optimization. Managing data in such applications is a dynamic and ever-evolving process, as data needs and user demands grow. By following best practices, adhering to data privacy regulations, and leveraging modern technologies and tools, developers can embark on a data-driven journey that ensures their complex mobile applications remain robust, responsive, and user-centric.

Advanced data storage solutions

Data is at the core of mobile app development, powering everything from user profiles and preferences to content and transactions. As mobile apps become increasingly complex and feature-rich, the demand for advanced data storage solutions has grown significantly. Developers must choose the proper data storage approach to ensure the app's performance, scalability, and data integrity. This section will explore the landscape of advanced data storage solutions in mobile app development, highlighting key considerations and best practices.

Relational databases, such as SQLite, MySQL, and PostgreSQL, have long been a mobile app data storage staple. These databases offer structured and efficient data storage and retrieval mechanisms, making them suitable for applications that require complex data relationships and querying capabilities. SQLite, in particular, is a popular choice for mobile apps due to its lightweight nature and native support on both Android and iOS platforms.

NoSQL databases have gained prominence in mobile app development, providing flexible and schema-less data storage options. Solutions like Firebase Realtime Database and MongoDB are popular choices. Firebase Realtime Database offers real-time data synchronization, making it suitable for collaborative and chat applications, while MongoDB's document-oriented structure accommodates diverse data types and evolving data schemas.

Cloud-based databases, such as Amazon DynamoDB, Google Cloud Firestore, and Microsoft Azure Cosmos DB, offer scalability and high availability. These databases are designed to handle data storage and access at scale, making them suitable for mobile apps with large user bases or varying workloads. Cloud-based solutions often provide global replication and low-latency access, ensuring a consistent user experience across regions.

Object storage solutions like Amazon S3 or Google Cloud Storage are essential for apps that handle large volumes of media files. These services enable efficient storage and retrieval of images, videos, and other media assets.
Mobile apps can offload the burden of media storage to these specialized services, ensuring optimal performance and reducing server-side complexities.

Key-value stores, such as Redis and LevelDB, are ideal for scenarios where high-speed data access is critical. These stores are used to cache frequently accessed data or manage session data in real-time applications. Key-value stores are lightweight and excel at read-heavy workloads, enhancing app responsiveness.

Data security is paramount in mobile app development, particularly for applications handling sensitive user information or confidential data. Developers can utilize encrypted storage solutions and secure containers to protect data at rest. Technologies like Apple's Data Protection API for iOS and Android's Encrypted File API provide robust encryption mechanisms to safeguard data.

GraphQL is an API query language that enables applications to request the exact data they require. Mobile apps can use GraphQL to fetch data from a backend server efficiently. Additionally, Backend as a Service (BaaS) providers like Firebase and AWS Amplify offer pre-built backend components for data storage, authentication, and more, simplifying backend development for mobile apps.

Offline data storage solutions, like Realm and Couchbase Mobile, enable apps to work seamlessly offline. These databases offer features like data synchronization, conflict resolution, and local caching, ensuring that users can access and interact with app data even when they are not connected to the internet.

Choosing the appropriate data storage solution is a foundational decision in mobile app development. The choice should align with the app's requirements, user expectations, and scalability needs. As apps become more complex, managing and optimizing data storage becomes increasingly critical. By embracing advanced data storage solutions and best practices, developers can lay a solid foundation for mobile apps that excel in performance, responsiveness, and user satisfaction.

Offline data synchronization

Mobile app users expect seamless experiences, whether connected to the internet or offline. Mobile app developers have turned to offline data synchronization techniques to meet these expectations. Offline data synchronization ensures that app data remains accessible and up-to-date even when a device is not connected to the internet. This capability has become increasingly vital as mobile apps become more complex and rely on real- time data. In this section, we will delve into the significance of offline data synchronization in mobile app development, its challenges, and best practices for implementation.

Offline data synchronization is critical because it allows users to interact with an app's data seamlessly, regardless of their network connection. Imagine a task management app: Users should be able to create, update, and view tasks whether they're in an area with a strong Wi-Fi signal or no signal at all. Offline synchronization ensures users can continue working without disruption.

Mobile networks can be unreliable, with users frequently moving in and out of connectivity zones. Offline synchronization reduces an app's dependency on a consistent network connection, improving the overall user experience. It eliminates the frustration of waiting for content to load or encountering error messages due to poor connectivity.

Offline-first strategies prioritize local data access and enhance an app's performance and responsiveness. By storing frequently accessed data locally, apps can provide near-instantaneous access, resulting in a more fluid and engaging user experience. This is particularly crucial in situations where real-time interaction is necessary, such as messaging apps or collaborative tools.

Synchronizing data offline introduces challenges related to data consistency and conflict resolution. When multiple devices or users modify the same data while offline, conflicts may arise when attempting to sync changes. Implementing conflict resolution strategies, such as last-write-wins or manual resolution, ensures data integrity and consistency.

To enable offline data synchronization, mobile apps employ various offline storage solutions. Local databases, such as SQLite for Android or CoreData for iOS, serve as repositories for locally cached data. These databases allow developers to store, query, and update data efficiently while offline. NoSQL databases like Firebase Realtime Database or Couchbase Mobile offer real-time synchronization capabilities for offline-first scenarios.

Developers employ different synchronization strategies based on their app's requirements. Push-based synchronization pushes local changes to a remote server when a network connection is available. Pull-based synchronization fetches updates from the server to the local device. Bidirectional synchronization combines both approaches to ensure data consistency across devices and servers.

Conflict resolution mechanisms are essential when handling conflicts arising from simultaneous data modifications. Apps can adopt various conflict resolution strategies, including timestamp-based resolution (the most recent change prevails), user-defined resolution (allowing users to decide), and automatic conflict resolution algorithms.

Offline data synchronization often involves caching and prefetching data to ensure smooth offline interactions. Apps can intelligently cache data based on user behavior and preferences, allowing users to access content they are likely to need while offline. Prefetching frequently accessed data ensures a seamless offline experience. Offline data synchronization is a game-changer in mobile app development, empowering apps to deliver consistent, responsive, and uninterrupted user experiences. Implementing effective offline synchronization strategies, storage solutions, and conflict resolution mechanisms is essential for ensuring data consistency and reliability in offline scenarios. As mobile apps continue to evolve and users' expectations rise, the ability to seamlessly handle offline data synchronization becomes not just a feature but a fundamental requirement for successful mobile applications. Developers who master these techniques can create apps that truly excel in providing a continuous and engaging user experience.

Data encryption and security best practices

In today's digital era, data security is crucial, particularly for mobile app developers who handle private content, financial information, and sensitive user data on a daily basis. Data breaches can have disastrous effects on users and app developers alike. To safeguard against these risks, implementing robust data encryption and security measures is not just good practice but an ethical and legal obligation. In this section, we will explore the significance of data encryption and security in mobile app development and delve into the best practices that developers should adhere to in order to protect user data and maintain trust.

End-to-end encryption (E2E) is the gold standard for securing data in transit. Data is encrypted on the device used by the sender and can solely be decrypted on the recipient's device due to E2E. Messaging apps like WhatsApp and Signal have demonstrated the effectiveness of E2E encryption in protecting user conversations from eavesdropping. Implementing E2E encryption using secure protocols like TLS/SSL or Signal's open-source encryption library is crucial for apps that transmit sensitive user data.

Data at rest encryption ensures that data stored on a device or server remains secure, even if an unauthorized entity obtains access to the physical storage medium. Developers should utilize platform-specific encryption libraries and mechanisms, such as Apple's Data Protection API for iOS or Android's Encrypted File API, to protect user data on the device. Additionally, strong encryption algorithms like AES (Advanced Encryption Standard) ensure that data remains unreadable without the appropriate decryption keys.

Authentication mechanisms, such as password-based authentication, biometrics, or multi-factor authentication (MFA), are essential for verifying the identity of app users. To prevent illegal access to user accounts and private information, developers should employ secure authentication procedures. Fine-grained authorization restrictions ought to be implemented as well to guarantee that users can only access the information and functionalities that they are permitted to utilize.

Effective encryption relies on secure key management. The keys that are used to encrypt and decrypt data must be kept safe and shielded from prying eyes. Mobile app developers should avoid hardcoding encryption keys in the app's source code and utilize secure key storage mechanisms provided by the platform, such as the iOS Keychain or Android Keystore.

The choice of communication protocols plays a significant role in data security. Developers should opt for secure communication protocols like HTTPS for web requests and TLS for secure socket connections. Avoiding insecure protocols like HTTP and unencrypted Wi-Fi connections is crucial for protecting data in transit.

Regular security audits and penetration testing are necessary for identifying and mitigating security vulnerabilities in mobile apps. Security audits involve reviewing code, architecture, and data handling processes to ensure compliance with security best practices. Penetration testing involves actively testing the app for vulnerabilities by attempting to exploit weaknesses. Both approaches help developers address potential security risks proactively.

Data privacy regulations, like GDPR and CCPA, impose strict requirements for collecting, storing, and processing user data. Mobile app developers must ensure compliance with these regulations, including obtaining user consent for data collection, providing data export options, and offering mechanisms for user data erasure. Compliance not only safeguards user rights but also builds trust.

A secure mobile app extends beyond the device itself. Developers should ensure that the backend infrastructure, including servers and databases, is also secure. Employing practices like firewalls, intrusion detection systems, and routine security updates helps protect the backend from threats.

In mobile app development, data encryption and security are not just features; they are foundational principles that uphold user trust and protect sensitive information. Mobile app developers must prioritize security best practices throughout the development lifecycle to safeguard user data from threats and breaches. As the digital landscape evolves and cyber threats become increasingly sophisticated, staying vigilant and proactive in implementing robust encryption and security measures is mobile app developers' ethical and responsible path. By following to these best practices, developers can create apps that excel in functionality and user experience and provide the highest level of data protection and security.

CHAPTER V

Advanced Networking and APIs

Working with advanced API integration

In the fast-paced world of mobile app development, accessing external data and services through APIs (Application Programming Interfaces) is essential for enhancing app functionality and providing users with real-time information. Basic API integration is common, but as mobile apps become more sophisticated, developers must master advanced API integration techniques to meet user expectations and create seamless, data-rich experiences. This section will explore the significance of advanced API integration in mobile app development, its challenges, and best practices for success.

Advanced API integration opens the door to enriching user experiences by pulling in real-time data, enabling features like geolocation, integrating social media sharing, or offering personalized recommendations. For instance, weather apps can access weather data from external APIs, e-commerce apps can display real-time product prices, and travel apps can provide up-to-the-minute flight information, all contributing to a more engaging user experience.

In apps that rely on user accounts or profiles, advanced API integration ensures that user data remains synchronized across devices and platforms. Users expect their data, such as preferences, bookmarks, and saved content, to be consistent whether they access the app on their mobile device, tablet, or desktop. Implementing synchronization through APIs facilitates this seamless experience.

Apps requiring real-time data updates, such as news, sports, and messaging apps, depend on advanced API integration to instantly provide users with the latest information. By integrating with push notification APIs or WebSockets, developers can notify users of new content, messages, or events in real-time, enhancing app engagement.

Advanced API integration empowers developers to tailor app experiences to individual users. Developers can create personalized user journeys through APIs that provide user-specific recommendations, content, or services. For example, streaming platforms use API integrations to recommend movies and music based on users' viewing habits.

While advanced API integration offers numerous benefits, it also presents challenges. Developers must handle issues such as rate limiting, API versioning, handling data inconsistencies, and managing authentication tokens securely. Additionally, developers must be prepared for API changes, downtime, or discontinuation and have contingency plans in place to minimize disruptions.

Security is paramount when working with APIs, especially in advanced integrations where sensitive user data may be exchanged. Developers should employ secure authentication methods, like OAuth 2.0, and implement encryption for data in transit. Additionally, developers must keep API keys and secrets secure and monitor API usage for potential security threats.

Robust error handling and resilience mechanisms are crucial for maintaining app stability during advanced API integration. Developers should anticipate and gracefully handle API errors, retries, and network issues. Proper error logging and user-friendly error messages also contribute to a smoother user experience.

Thorough testing and continuous monitoring are essential to ensure that advanced API integrations function correctly and reliably. Developers should employ testing frameworks, conduct load testing, and employ API mocking tools during development. Post-launch, monitoring tools can track API performance, detect anomalies, and provide insights for optimization.

Maintaining version control of APIs is vital to ensure compatibility and avoid unexpected disruptions. Developers should rely on well-documented APIs that provide clear guidelines, usage examples, and versioning information. API documentation helps developers understand endpoints, parameters, and expected responses.

Advanced API integration is the cornerstone of modern mobile app development, enabling developers to create dynamic, data-driven, and highly personalized experiences. By embracing advanced integration techniques, developers can enhance user engagement, provide real-time updates, and access a wealth of external data and services. However, these benefits come with the responsibility of addressing challenges, ensuring security, and prioritizing error handling and resilience. As mobile apps continue to evolve and users demand richer, more interactive experiences, mastering advanced API integration becomes essential for staying competitive and delivering apps that meet the highest functionality and user satisfaction standards.

Handling authentication and authorization

Authentication and authorization are integral to mobile app development, safeguarding user data and ensuring users access only the resources and features they are entitled to. Developers must prioritize robust and secure authentication and authorization mechanisms in an age of escalating data breaches and privacy concerns. In this section, we will delve into the significance of handling authentication and authorization in mobile app development, the challenges they present, and best practices for their implementation.

Verifying the identity of a user or other entity attempting to access an application or system is known as authentication. User authentication is crucial to mobile app development since it guarantees that only authorized users may access the functionality and data of the app. Common authentication methods include username and password, biometrics (fingerprint, face recognition), email verification, and social login via third-party platforms like Google or Facebook.

Authorization defines what actions authenticated users are allowed to perform within the app. It determines the user's access level to specific resources, functionalities, or data. Authorization is often implemented through roles and permissions, where each user is assigned a role with associated permissions. For instance, a role might grant a user read-only access to specific data but restrict their ability to modify it.

Securely storing user data, including login credentials, is paramount in mobile app development. Developers should avoid storing sensitive information in plaintext and instead use encryption and hashing techniques to protect passwords and other confidential data. Mobile platforms offer secure storage options like the iOS Keychain and Android Keystore for managing sensitive information.

OAuth and OpenID Connect are widely adopted mobile app user authentication protocols. OAuth enables secure third-party access to resources without exposing user credentials. OpenID Connect builds on OAuth to provide identity verification and authentication services, making it suitable for scenarios where user identity is critical. Because they are safe and convenient, biometric authentication techniques like fingerprint and facial recognition are becoming more and more utilized. Mobile apps can leverage biometric authentication APIs from platforms like Android's BiometricPrompt and iOS's Touch ID and Face ID.

Users can access numerous services or apps with just one authentication and save time by not having to enter their credentials again due to Single Sign-On (SSO). Implementing SSO through technologies like JSON Web Tokens (JWT) or SAML (Security Assertion Markup Language) streamlines the user experience while maintaining security.

Two-Factor Authentication (2FA) adds a further layer of security by asking users to provide two separate authentication factors, typically something they know (e.g., a password) and something they have (e.g., a mobile device or authentication app). Mobile apps should offer 2FA as an optional but highly recommended security feature.

Effective session management is crucial to maintaining user authentication during a user's session. Developers must implement secure session handling mechanisms to prevent session hijacking and ensure that sessions expire after a reasonable period of inactivity.

Role-Based Access Control (RBAC) is a widely adopted authorization model that categorizes users into roles and assigns permissions based on those roles. Mobile apps can implement RBAC to ensure that users have access only to the features and data relevant to their roles, enhancing security and data privacy.

When mobile apps communicate with backend services or APIs, secure API authentication is essential. Developers should use technologies like API keys, OAuth tokens, or JWTs to verify the authenticity of API requests and prevent unauthorized access to server resources.

In the era of mobile app development, user data protection is non-negotiable. Authentication and authorization mechanisms are the guardians of this data, ensuring that only authorized users access it and that they do so securely. Developers must master these techniques, leveraging secure storage, industry-standard protocols like OAuth and OpenID Connect, and modern authentication methods like biometrics and 2FA. By prioritizing robust authentication and authorization, mobile app developers protect user data and build and maintain trust, an invaluable asset in today's data-centric landscape.

Implementing RESTful and GraphQL APIs

In the ever-evolving mobile app development landscape, efficient and flexible data retrieval and manipulation are crucial. To meet these demands, developers often turn to APIs (Application Programming Interfaces) to access and interact with external services and data sources. RESTful

(Representational State Transfer) and GraphQL are two prominent approaches for building APIs. In this section, we will explore the significance of implementing RESTful and GraphQL APIs in mobile app development, comparing their characteristics, use cases, and best practices for integration.

RESTful APIs adhere to the principles of REST, a set of architectural guidelines for designing networked applications. They rely on standard HTTP methods (GET, POST, PUT, DELETE) and use resource-based URLs to represent data and actions. RESTful APIs provide a clear and organized structure, making them easy to understand and work with. They are well-suited for scenarios where data retrieval follows a predetermined pattern, such as retrieving user profiles, product listings, or blog posts.

GraphQL is an API query language that enables clients to request the exact data they require. Unlike REST, which exposes fixed endpoints with predefined data structures, GraphQL enables clients to define their data requirements in a single query. This flexibility empowers mobile app developers to fetch only the necessary data, reducing over-fetching and under-fetching of information. GraphQL is ideal for scenarios where the data requirements may vary, such as in complex mobile apps with varying content or personalized user experiences.

One of the primary advantages of GraphQL is its ability to mitigate the issues of over-fetching and under-fetching data. In RESTful APIs, clients often receive more data than they need, resulting in wasted bandwidth and slower app performance. In contrast, GraphQL allows clients to request only the specific fields and data they require, optimizing data transfer efficiency.

RESTful APIs may require versioning when changes are made to the API structure or behavior to maintain compatibility with existing clients. This can lead to endpoint proliferation and complexities in handling different versions. GraphQL, on the other hand, allows for incremental changes without needing versioning, as clients specify their data requirements in the query, and the server responds accordingly.

GraphQL excels in handling complex queries and nested data structures. Mobile apps often require fetching data from multiple sources or aggregating data from different endpoints. GraphQL simplifies this process by allowing clients to define a single query encompassing all necessary data, reducing the need for multiple API calls. RESTful and GraphQL APIs can benefit from caching mechanisms to reduce redundant network requests. However, GraphQL's fine-grained control over data fetching enables more precise caching strategies. GraphQL's subscription model also allows for real-time updates, making it suitable for apps requiring live data, such as messaging or collaborative tools.

RESTful APIs often rely on documentation to describe available endpoints and their parameters. While well-documented REST APIs are essential, GraphQL takes this a step further with introspection capabilities that allow clients to explore the API's schema dynamically. This self-documentation feature simplifies client development and testing.

Adhering to best practices is crucial to implementing RESTful or GraphQL APIs in mobile app development. For RESTful APIs, maintain a consistent resource structure, use appropriate HTTP methods, and version your APIs when necessary. For GraphQL, design a clear schema, optimize queries, and handle security and authentication effectively.

The choice between RESTful and GraphQL APIs in mobile app development depends on the app's specific requirements and data access patterns. RESTful APIs offer simplicity and well-defined endpoints, making them suitable for straightforward data retrieval scenarios. In contrast, GraphQL provides flexibility, efficiency, and real-time capabilities, making it ideal for complex mobile apps with varying data needs. By understanding the strengths and trade-offs of each approach and adhering to best practices, mobile app developers can implement the most suitable API type to meet their app's demands and deliver a seamless and efficient user experience.

Advanced error handling and retry mechanisms

Our daily lives are now dramatically impacted by mobile apps, offering various functionalities from communication to entertainment and productivity. In this context, robust error handling and retry mechanisms are paramount. Advanced error handling goes beyond displaying a generic error message; it involves identifying, categorizing, and responding to errors intelligently. Retry mechanisms ensure that the app gracefully handles transient issues, such as poor network connectivity, without disrupting the user experience. This section will explore the significance of advanced error handling and retry mechanisms in mobile app development, their challenges, and best practices for their implementation.

Effective error handling is synonymous with a positive user experience. When errors occur, users appreciate informative and user-friendly error messages that guide them on resolving the issue or taking appropriate action.

Advanced error handling techniques ensure that users are not left frustrated but are instead provided with meaningful feedback and assistance.

Not all errors are equal, and categorizing them based on their severity and impact is essential. By classifying errors into categories like validation, server, or network errors, developers can tailor error messages and responses accordingly. For example, a validation error might prompt users to correct input fields, while a server error should be reported to the development team for investigation.

Mobile apps often encounter transient errors, such as network timeouts or server unavailability. Implementing retry mechanisms allows apps to automatically attempt the same operation again after a brief delay, improving the chances of success. Developers should carefully design retry logic to avoid overloading servers with repeated requests and provide clear feedback to users during retries.

Advanced error handling includes comprehensive error logging and reporting. Logging errors with relevant details, such as timestamps and user actions leading to the error, assists developers in diagnosing and resolving issues promptly. Real-time error reporting tools, like Crashlytics and Sentry, enable developers to monitor app performance and identify recurring errors.

For apps that rely on external services or APIs, gracefully degrading functionality in the face of errors is a best practice. For example, an e-commerce app could display cached product listings when unable to fetch real-time data from the server. This ensures that users can still access essential app features even during temporary service disruptions.

Mobile apps should provide users with avenues for reporting errors and seeking assistance. Implementing feedback mechanisms, such as in-app feedback forms or direct links to customer support, encourages users to communicate issues they encounter. Responsiveness to user feedback fosters a sense of trust and engagement.

Thorough testing, including simulated error scenarios, is essential for validating error handling and retry mechanisms. Developers can use testing tools and frameworks to simulate various error conditions, ensuring the app responds appropriately. Testing should encompass functional and stress testing to assess the app's performance under load.

In a global market, providing error messages in the user's preferred language is considerate and user-centric. Localized error messages ensure that users from diverse linguistic backgrounds can understand and act upon the information provided.

Advanced error handling is an ongoing process. Mobile apps should be designed for iterative development, with regular updates that address user-reported issues and improve error handling based on real-world usage patterns.

While error handling primarily focuses on user experience, developers must also consider security implications. Avoid disclosing sensitive information in error messages, as attackers can exploit such details. Error messages should be informative without compromising security.

In mobile app development, advanced error handling and retry mechanisms are not just technical considerations but integral to building trust and maintaining user engagement. Users expect apps to work reliably and to provide meaningful responses when issues arise. By categorizing errors, implementing retry strategies, and prioritizing user feedback and support, developers can ensure that their apps offer a seamless and user-friendly experience, even in the face of occasional errors. Effective error handling is not a sign of an app's flaws but a testament to its commitment to delivering a superior user experience.

CHAPTER VI

Advanced Performance Optimization

Profiling and optimizing app performance

In the highly competitive field of mobile app development, performance is a crucial factor that can make or break an app's success. Users expect fast, responsive, and efficient experiences, and any perceived sluggishness can lead to frustration and abandonment. Profiling and optimizing app performance are essential steps in ensuring that an app meets these high expectations. This section will explore the significance of profiling and optimizing app performance, the challenges they present, and best practices for achieving peak performance.

Instant gratification is the norm in this day and age, users have little patience for slow or unresponsive apps. They expect apps to load quickly, respond to their interactions without delay, and consume minimal device resources. Meeting these expectations is not just a matter of delighting users but also retaining them.
Profiling involves the systematic measurement and analysis of an app's performance characteristics.

Developers use profiling tools and techniques to identify bottlenecks, memory leaks, CPU usage, and other performance-related issues. Profiling provides actionable insights into areas that require optimization.
Profiling typically focuses on key performance metrics, including app launch time, screen load times, rendering

speed, and response to user input. These metrics help developers pinpoint areas of concern and prioritize optimization efforts.

Memory management is critical for app performance. Memory leaks, where unused memory is not released, can lead to performance degradation and eventual crashes. Profiling helps identify memory leaks and excessive memory usage, enabling developers to optimize memory management.

Apps that rely on network requests must optimize their network performance. Profiling network requests reveals latency issues, slow APIs, or excessive data transfer, allowing developers to implement optimizations like data compression, request batching, or caching.

UI rendering performance directly impacts the app's responsiveness. Profiling tools can highlight rendering bottlenecks, such as excessive UI updates or inefficient layout calculations. Optimizations may involve reducing unnecessary UI updates and optimizing layout algorithms.

Apps that drain device battery or consume excessive CPU resources face user backlash. Profiling reveals resource-intensive operations, enabling developers to implement energy-efficient coding practices and optimize CPU usage for a smoother user experience.

Once profiling identifies performance bottlenecks, developers can implement various optimization techniques. These may include code refactoring, lazy loading of resources, background processing, image compression, and adopting efficient algorithms and data structures.

Device fragmentation is a challenge in mobile app development. Apps must perform well on various devices with varying hardware capabilities. Profiling on different devices and configurations helps ensure consistent performance across the user base.

Optimizing app performance is not a one-time task. Developers should continuously monitor and profile the app, especially after updates or changes. Performance regressions can occur, and proactive monitoring helps identify and address them promptly.

User feedback is a valuable source of insights into app performance issues. Developers should encourage users to report performance-related problems and use this feedback to guide optimization efforts.

A/B testing can help assess the impact of performance optimizations on user engagement and retention. By comparing the performance-optimized version of the app with the previous version, developers can make data-driven decisions about the effectiveness of their efforts.

In the fiercely competitive world of mobile app development, performance is not just a feature but a competitive advantage. Profiling and optimizing app performance are essential for meeting user expectations, retaining users, and ensuring an app's success. By systematically identifying bottlenecks, addressing memory issues, optimizing network requests, and implementing resource-efficient coding practices, developers can create apps that function flawlessly and provide a responsive and enjoyable user experience. In an ecosystem where user satisfaction is paramount, prioritizing performance optimization is a commitment to building apps that stand out and thrive.

Techniques for reducing app size

In the ever-evolving landscape of mobile app development, delivering a compact and efficient app is critical to success. Users increasingly demand apps that are not only feature-rich but also lightweight and quick to download. Reducing app size has numerous advantages, from improving user experience to minimizing data usage and conserving device storage. In this section, we will explore the significance of reducing app size, the challenges developers face in this endeavor, and a comprehensive set of techniques and best practices for achieving a lean and efficient mobile app.

App size matters in today's mobile ecosystem. Users are mindful of their data plans and device storage, and large apps can deter downloads and updates. A smaller app size not only enhances user experience by reducing installation and update times but also lowers data consumption and conserves device storage space. While the benefits of reducing app size are clear, developers face several challenges in achieving this goal. Balancing features and functionality with size constraints can be a delicate dance. Additionally, developers must consider platform-specific limitations, optimize app assets, and ensure that the reduced size does not compromise app performance or user experience.

Images and multimedia assets often constitute a significant portion of an app's size. Developers should employ image compression techniques to reduce file sizes without sacrificing visual quality. Using modern image formats like WebP or AVIF can also help achieve smaller file sizes. Furthermore, employing vector graphics for icons and images reduces the need for multiple resolutions, further shrinking the app's footprint.

Some app components can be delayed in loading until they are actually needed by using a technique called lazy loading. For example, images below the fold in a scrollable view can be loaded only when they come into view. Similarly, on-demand resources allow apps to download and install additional content when required, minimizing the initial download size.

Modularization involves breaking down an app into smaller, self-contained modules. This enables users to download only the essential components when first installing the app and fetch additional modules as needed. App bundles, available on Android, facilitate modularization by delivering optimized and smaller packages to users.

Proguard is a tool for Android apps that performs code shrinking and obfuscation, reducing the size of the compiled code. Similarly, iOS developers can employ tree shaking, a technique that removes unused code and resources from the app bundle, resulting in a smaller app size.

Apple's App Thinning is a suite of technologies that optimize the installation process and reduce the app's size on iOS devices. It includes on-demand resources, slicing (delivering only the necessary app assets for the user's device), and bitcode compilation (allowing the App Store to re-optimize the app for specific devices).

Compressing app assets and resources, such as audio and video files, can significantly reduce the app's size. Streaming assets, where portions of media content are fetched as needed during playback, help minimize initial download size while providing a seamless user experience.

Code splitting involves breaking the app's JavaScript code into smaller modules that are loaded as needed. This technique is commonly used in web apps and can be adapted to mobile app development. Additionally, minification tools reduce code size by removing unnecessary characters and whitespace.

Over time, app dependencies can accumulate, some of which may no longer be necessary. Regularly auditing and removing unused libraries and dependencies can significantly reduce app size.

For some use cases, Progressive Web Apps (PWAs) offer a lightweight alternative to traditional mobile apps. PWAs are web apps that offer a native-like experience and may be installed on a user's device. They often have smaller footprints since they leverage web technologies.

Apps that store user-generated content, such as media uploads, can optimize storage by compressing or resizing user data. Implementing caching and eviction policies can also help manage local data storage efficiently.

Listening to user feedback and monitoring app usage can provide insights into areas where size optimization is needed. Users often report concerns about app size, and their feedback can guide developers in making targeted improvements.

A/B testing allows developers to assess the impact of size optimizations on user engagement and retention. By comparing the performance of a smaller version of the app with the original, developers can make data-driven decisions about the effectiveness of their efforts.

Reducing app size is a delicate balance between providing valuable features and delivering a lightweight user

experience. Developers must continuously optimize their apps to meet user expectations for fast downloads, minimal data usage, and efficient device storage utilization. By employing a combination of image optimization, modularization, code shrinking, and other size-reduction techniques, developers can create mobile apps that are lean and efficient and deliver a seamless and enjoyable user experience. In today's mobile landscape, where users expect responsiveness and efficiency, app size optimization is not just a best practice
but a competitive advantage.

Advanced memory management

Memory management is a critical aspect of mobile app development that can profoundly impact an app's efficiency and performance. Mobile devices come with limited memory resources, and poorly managed memory can lead to sluggish performance, crashes, and frustrated users. Advanced memory management techniques are essential for optimizing memory usage, preventing memory-related issues, and delivering a seamless user experience. This section will explore the significance of advanced memory management in mobile app development, its challenges, and best practices for efficient memory handling.

Efficient memory management ensures a mobile app runs smoothly and delivers a responsive user experience. Inefficient memory usage can lead to problems like slow app launch times, unresponsiveness during usage, and even app crashes. Since users have high expectations for app performance, effective memory management is a fundamental requirement.

Mobile app developers face several challenges in managing memory effectively. Mobile devices have limited physical memory; apps must compete for resources with the operating system and other running applications. Additionally, the diversity of devices and platforms introduces complexity in memory management, as different devices have varying memory capacities and behaviors.

Modern mobile platforms, like iOS and Android, incorporate automatic memory management systems that assist developers in managing memory. On iOS, Automatic Reference Counting (ARC) automatically tracks and releases memory when objects are no longer used. Android's Garbage Collector (GC) serves a similar purpose, reclaiming memory occupied by no longer reachable objects.

Choosing the right data structures can significantly impact memory usage. Developers should select memory-efficient data structures for the specific tasks they need to perform. For example, using sparse data structures for sparse data sets can reduce memory overhead.

One method that delays loading some resources or data until they are truly needed is called lazy loading. For instance, images can be loaded only when the user scrolls to them in a list of items, reducing the app's initial memory footprint.

For apps that deal with large amounts of data, offloading less frequently used data to disk storage rather than keeping it all in memory can be a memory-saving strategy. Caching data on disk and retrieving it as needed can improve performance without consuming excessive memory.

Mobile apps should prioritize resources based on their importance and relevance. For example, resources related to the current screen or user interactions should be kept in memory, while resources related to background tasks can be released or offloaded to disk.

Memory profiling tools and techniques help developers identify memory leaks and inefficient memory usage patterns. Profiling enables developers to pinpoint the sources of memory issues and take corrective actions.

Memory pools are pre-allocated blocks of memory that can be reused for specific tasks, reducing memory allocation overhead. By recycling memory from pools, developers can minimize memory fragmentation and improve memory efficiency.

Comprehensive testing, including stress testing and memory leak detection, is essential to ensure an app's robust memory management. Automated testing tools can help identify memory-related issues during development.

Memory management is an ongoing process. Developers should continuously monitor an app's memory usage, especially after updates or changes. Monitoring helps detect memory regressions and ensures that the app remains memory-efficient.

Advanced memory management is not an optional aspect of mobile app development but a cornerstone of app performance. In an environment where users demand speedy and responsive experiences, efficient memory handling is a non-negotiable requirement. By employing automatic memory management systems, choosing memory-efficient data structures, implementing lazy loading, and conducting thorough testing, developers can optimize memory usage, prevent memory-related issues, and deliver apps that meet user expectations. Memory management may be a behind-the-scenes aspect of app development, but its impact on user satisfaction and app success is undeniable. In the mobile app development world, where performance is vital, advanced memory management is a skill that every developer must master.

Multithreading and concurrency in mobile apps

Multithreading and concurrency are powerful techniques that can significantly enhance the performance and responsiveness of mobile apps. In the dynamic world of mobile app development, where users expect seamless and snappy experiences, these techniques are essential for leveraging the full potential of modern mobile devices. This section will explore the significance of multithreading and concurrency in mobile apps, their challenges, and best practices for harnessing their capabilities.

Mobile devices boast impressive hardware capabilities today, with multicore processors becoming the norm. Multithreading and concurrency allow developers to take full advantage of this hardware, enabling apps to perform multiple tasks simultaneously. This results in faster execution, smoother animations, improved user interface responsiveness, and the ability to perform resource-intensive operations without freezing the app.

While multithreading and concurrency offer substantial benefits, they also introduce challenges that developers must address. Concurrent access to shared resources can result to data corruption and synchronization issues. Deadlocks and race conditions are common pitfalls in multithreaded programming that can result in app crashes and erratic behavior. Developers need to design and manage concurrent processes to mitigate these challenges carefully.

In mobile app development, the main thread, also known as the UI thread, is responsible for handling user interface operations. If long-running tasks, such as network requests or complex computations, are executed on the main thread, they can cause the UI to freeze, leading to a poor user experience. Multithreading lets developers offload such tasks to background threads, ensuring the UI remains responsive.

Asynchronous programming is a fundamental aspect of concurrency in mobile apps. It allows developers to initiate tasks and continue with other operations without waiting for the tasks to complete. Popular asynchronous programming techniques include callbacks, promises, and async/await, which simplify managing asynchronous tasks and handling their results.

Thread pools are a common concurrency mechanism that helps manage and reuse threads efficiently. By maintaining a pool of worker threads, apps can avoid the overhead of thread creation and destruction. Thread pools are beneficial when dealing with a high volume of short-lived tasks.

iOS developers can access Grand Central Dispatch (GCD), a powerful framework for managing multithreading and concurrency. GCD abstracts the complexities of thread management, offering a high-level API for dispatching tasks asynchronously. Developers can leverage GCD's queues and dispatch groups to effectively organize and control concurrent operations.

Android provides AsyncTask and Executor frameworks to simplify multithreading. AsyncTask is designed for simple background tasks, while Executors offer more fine- grained control over thread management. Developers can choose the appropriate framework based on the complexity of their concurrency needs.

To prevent data corruption in multithreaded apps, synchronization mechanisms such as locks, semaphores, and mutexes are employed. These tools ensure that only one thread can access a shared resource at a time, preventing conflicts and maintaining data integrity.

Designing for thread safety is essential in concurrent programming. Immutable data structures, which cannot be modified after creation, are inherently thread-safe and simplify multithreaded code. By minimizing shared mutable state, developers can reduce the risk of concurrency-related bugs.

Multithreaded code can be challenging to test and debug, as issues may be non-deterministic and hard to reproduce. Comprehensive testing, including stress testing and race condition detection, ensures that multithreaded apps behave correctly under varying conditions.

In the fast-paced world of mobile app development, where performance and responsiveness are paramount, multithreading and concurrency are indispensable tools. They empower developers to fully utilize the capabilities of modern mobile devices and deliver apps that respond quickly, handle resource-intensive tasks seamlessly, and provide a superior user experience. While synchronization and data sharing challenges in multithreaded environments require careful consideration and robust programming techniques, the benefits far outweigh the complexities. Multithreading and concurrency have become essential skills for mobile app developers looking to create high-performance, responsive, and user-friendly apps that thrive in today's competitive landscape.

CHAPTER VII

Advanced Testing and Quality Assurance

Implementing effective testing strategies

Testing is a cornerstone of mobile app development, playing a pivotal role in ensuring that apps are of high quality, free from critical defects, and capable of providing a smooth user experience. Effective testing strategies are essential in the fast-paced world of mobile app development, where competition is fierce, and users have high expectations. This section will explore the significance of implementing effective testing strategies in mobile apps, the challenges they present, and best practices for achieving reliable and robust testing outcomes.

Mobile apps are integral to modern life, serving as tools for communication, entertainment, productivity, and more. Users expect apps to work flawlessly, and any issues, from crashes to poor performance, can result in negative reviews, uninstalls, and lost opportunities. Effective testing is vital for identifying and rectifying issues before they reach users, ensuring that apps meet their expectations for quality and reliability.

Mobile app testing poses several challenges for developers and testers. The sheer diversity of devices, operating systems, screen sizes, and network conditions demands extensive testing coverage. Moreover, mobile apps often rely on external services and APIs, introducing additional complexities in testing integration points and data exchange. The rapid release cycles common in mobile app development also require efficient and automated testing approaches to keep up with the pace of updates.

Effective testing strategies encompass a variety of testing types, including functional testing, usability testing, performance testing, security testing, and more. Each type of testing addresses specific aspects of app quality and reliability. Functional testing ensures that app features work as intended, while usability testing evaluates user-friendliness and user interface design. Performance testing assesses app speed and resource usage, and security testing identifies vulnerabilities and ensures data protection.

Manual and automated testing have their roles in mobile app testing. Manual testing relies on human testers to execute test cases and evaluate app behavior. It is valuable for exploratory testing, usability assessments, and user experience evaluation. On the other hand, Automated testing, entails using scripts and testing frameworks to execute repetitive and systematic test cases efficiently. Automated testing is particularly beneficial for regression testing, ensuring new updates do not introduce regressions into previously working functionality.

Implementing continuous integration (CI) and continuous testing (CT) is crucial to keep pace with the rapid development cycles of mobile apps. CI ensures that code changes are integrated into a shared repository and built automatically, while CT involves running automated tests on the codebase continuously. This approach helps identify issues early, allowing developers to address them promptly and maintain app stability.

Testing on various devices and platforms is essential to ensure app compatibility. Testers should consider multiple device models, operating system versions, screen sizes, and network conditions to identify potential issues and guarantee that the app functions consistently across diverse user environments.

Real user testing, often conducted through beta testing programs, provides valuable insights into how users interact with the app in real-world scenarios. Beta testers can identify usability issues, provide feedback on app features, and report unexpected behavior that may not be evident in controlled testing environments.

Leveraging test automation frameworks can streamline the testing process and improve efficiency. Popular mobile app test automation frameworks like Appium, XCTest, Espresso, and Detox enable developers to create and execute test scripts across multiple platforms and devices.

Effective test data management ensures that testing environments accurately replicate real-world scenarios. Testers should carefully manage test data to cover various use cases and edge conditions, ensuring comprehensive test coverage.

Establishing a clear reporting mechanism and feedback loop for test results is essential for communication between developers and testers. Timely reporting of issues, coupled with clear documentation, helps developers understand and address problems efficiently.

Effective testing strategies are the foundation of quality assurance in mobile app development. They enable developers to identify and rectify issues, ensure that apps meet user expectations for quality and reliability, and maintain a competitive edge in the mobile app market. While mobile app testing presents challenges due to the diversity of devices and platforms, as well as the need for rapid releases, adopting a comprehensive approach that entails manual and automated testing, continuous integration, device coverage, and real user feedback ensures that apps are thoroughly tested and ready for prime time. In an ecosystem where user satisfaction is paramount, effective testing is not just a best practice but a prerequisite for app success.

Unit testing, integration testing, and UI testing

Quality assurance is a paramount aspect of mobile app development. Ensuring that an app functions correctly, performs well, and provides a seamless user experience requires a systematic and multifaceted approach to testing. Unit testing, integration testing, and UI testing are three essential components of this approach, each serving a unique purpose in the quest for app reliability and quality. This section will explore the significance of unit testing, integration testing, and UI testing in mobile app development, their respective roles, and how they work together to form a robust testing strategy.

Unit testing is the bedrock of software testing, and it focuses on the smallest testable units of an app's code, typically individual functions or methods. In mobile app development, unit tests assess the correctness of code at the most granular level, ensuring that each function performs its intended task accurately. Unit tests are designed to be fast, isolated, and repeatable, making them invaluable for catching and fixing bugs early in development.

Unit testing offers several advantages. It provides rapid feedback to developers, helping them identify and solve issues as they write code. Unit tests act as a safety net, preventing regressions by verifying that code changes do not break existing functionality. They also encourage modular and maintainable code by promoting code separation and encapsulation.

While unit tests focus on individual code units, integration testing zooms out to examine how various app components interact. Integration tests evaluate the connections between different modules, subsystems, or services, ensuring they work harmoniously. Integration testing verifies that components like APIs, databases, and external services collaborate in mobile apps as expected.

Integration testing identifies issues that may not be apparent during unit testing. It ensures that data flows smoothly between app components and that integrations with external services or third-party libraries are robust. By addressing integration issues early, developers can avoid complex debugging later in the development cycle.

UI testing, also known as end-to-end testing or user interface testing, evaluates an app's user interface and interactions from the user's perspective. In mobile apps, UI testing simulates user actions, such as tapping buttons and entering text, to verify that the app responds correctly. UI tests assess not only the visual aspects but also the functionality and user experience of the app.

UI testing provides a comprehensive view of the app's behavior, including how users will interact with it. By automating UI tests, developers can catch user interface issues, navigation problems, and usability concerns. UI tests help ensure the app meets user expectations for ease of use and responsiveness.

Unit testing, integration testing, and UI testing are often visualized as a testing pyramid, with unit tests forming the base, integration tests in the middle, and UI tests at the top. This pyramid illustrates the ideal testing distribution, where most tests are fast and inexpensive unit tests, followed by fewer integration tests and a smaller number of UI tests.

Effective quality assurance requires a holistic approach that combines unit testing, integration testing, and UI testing. Unit tests catch code-level issues, integration tests verify component interactions, and UI tests ensure the app functions as expected from the user's perspective. These testing layers offer comprehensive coverage, helping developers identify and rectify issues at every level of the app's architecture.

While each testing type has its merits, there are challenges to overcome. UI testing can be slow and fragile, especially as apps evolve. Developers must choose the right testing tools and frameworks and strike a balance between the number of tests and testing time. Additionally, maintaining a balance between unit, integration, and UI testing can be challenging, as over-reliance on one type can lead to gaps in test coverage.

Unit testing, integration testing, and UI testing are essential components of a comprehensive testing strategy for mobile apps. While each type serves a distinct purpose, they work in tandem to ensure that apps are free from critical defects, perform well, and provide a user-friendly experience. In an environment where user satisfaction and app quality are paramount, developers must embrace all three testing layers to create reliable and robust mobile applications. By doing so, they can confidently deliver apps that meet user expectations and excel in a competitive marketplace.

Continuous integration and continuous delivery (CI/CD)

The realm of mobile app development is marked by its rapid pace, ever-evolving user expectations, and intense competition. In such a dynamic environment, creating feature-rich, user-friendly mobile apps is crucial, but equally important is the need to deliver these apps efficiently and with uncompromising quality. Continuous Integration and Continuous Delivery are indispensable practices to address these challenges. This section delves into the world of CI/CD in mobile app development, shedding light on its principles, enumerating its benefits, and discussing best practices, highlighting its transformative impact on the app development process.

The mobile app market is fiercely competitive, with users demanding frequent updates, new features, and prompt bug fixes. In this context, traditional development approaches often struggle to meet the expectations for speed and quality. CI/CD serves as the solution, offering a systematic and automated approach to developing, testing, and deploying mobile apps.

Continuous Integration (CI) is the first pillar of the CI/CD pipeline. It involves frequently integrating code changes from multiple developers into a shared repository. With CI, developers can automate the process of building and testing code as soon as it's committed. This practice catches bugs and integration issues early in the development cycle, reducing the time and effort required to fix them.

Continuous Delivery (CD) complements CI by automating the deployment process. It ensures that each code change that passes the CI stage is automatically deployed to a staging or production environment. CD also entails running additional tests, such as user acceptance testing (UAT) and performance testing, to guarantee that the app is ready for release.

The benefits of CI/CD in mobile app development are substantial. It expedites the development cycle, enabling developers to release updates and new features more frequently, which is essential in a market where user preferences evolve swiftly. Furthermore, it elevates the quality of apps by automating testing and providing continuous feedback, minimizing the chances of bugs reaching production. Collaboration among developers, testers, and other stakeholders is also fostered, thanks to the unified and automated code integration, testing, and deployment pipeline. CI/CD minimizes the risk of human error and ensures that code changes are meticulously validated before reaching users, enhancing the overall efficiency of resource utilization.

To harness the full potential of CI/CD in mobile app development, adhering to best practices is essential. Developers should employ a version control system, such as Git, to manage code changes and revisions. This practice ensures the traceability and reversibility of code changes. Establishing a robust suite of automated tests, including unit tests, integration tests, and end-to-end tests, is pivotal. These tests should run automatically during the CI/CD pipeline.

Furthermore, CI/CD pipelines should produce immutable build artifacts, guaranteeing consistency and reproducibility in the deployment process. Containerization technologies, like Docker, are invaluable for packaging apps and their dependencies, ensuring consistent deployment across different environments. Implementing monitoring tools for real-time app performance tracking and user behavior completes the feedback loop, informing future development efforts. Security scanning tools must be integrated into the CI/CD pipeline to detect vulnerabilities in code and dependencies. While automated testing is crucial, manual testing should not be neglected, as human testers can provide essential insights into user experience and edge cases.

It is paramount to adopt a culture of ongoing learning and development. Regularly reviewing and updating the CI/CD pipeline to incorporate lessons learned and emerging best practices to ensure its continued effectiveness. Challenges may arise during the transition to CI/CD, necessitating changes in development processes, culture, and tooling. Developers must carefully consider factors such as test coverage, compatibility with legacy systems, and the cost of infrastructure and tools.

In conclusion, Continuous Integration and Continuous Delivery have revolutionized mobile app development, offering a systematic and automated approach that streamlines processes, enhances quality, and accelerates time to market. In a fiercely competitive landscape where user expectations evolve continuously, CI/CD is the path to success. By adhering to best practices, addressing challenges, and cultivating a culture of continuous improvement, mobile app developers can harness the power of CI/CD to create and deliver high-quality apps that meet user demands and stand out in the market. As the mobile app ecosystem evolves, CI/CD remains an indispensable practice, enabling developers to navigate the difficulties and seize the opportunities of the digital age.

App testing on different devices and platforms

The diverse landscape of mobile devices and platforms presents opportunities and challenges for app developers. With many devices running various operating systems and versions, thorough testing across this spectrum is crucial to ensure app compatibility, functionality, and quality. This section will explore the significance of app testing on different devices and platforms, the complexities it introduces, and best practices for achieving comprehensive compatibility and quality
assurance.

Mobile app development has revolutionized how we interact with technology, but this revolution has also brought complexity. Users today access apps on a wide range of devices, including smartphones, tablets, wearables, and more. These devices run on different operating systems, such as iOS, Android, and others, each with multiple versions and configurations. To provide a seamless user experience and reach a broad audience, it is imperative that apps function correctly on this diverse array of devices and platforms.
Android, in particular, is known for its device fragmentation. The Android ecosystem encompasses many device manufacturers, screen sizes, resolutions, hardware capabilities, and Android versions. Testing on various Android devices is essential to ensure that an app

looks and performs consistently across this diverse landscape.

Although iOS devices are more controlled regarding hardware and operating system versions, they still have significant variability. Testing on various iPhone and iPad models is necessary to confirm that an app behaves as expected and adapts to different screen sizes and resolutions.

Developers often turn to cross-platform development frameworks like React Native, Flutter, or Xamarin to streamline app development for multiple platforms. While these frameworks promise code reusability, thorough testing remains essential to account for platform-specific behaviors and differences in the user interface.

Testing on different devices and platforms involves a range of considerations. These include screen size and resolution, hardware capabilities (such as camera quality and sensors), platform-specific features (like iOS's Face ID or Android's fingerprint authentication), and performance characteristics (such as CPU and memory usage).

Emulators and simulators are invaluable tools for device and platform testing. They allow developers to replicate the behavior of various devices and operating systems on a single machine. While they are useful for early testing and development, it's crucial to complement emulator testing with real-device testing to account for nuances that emulators may not capture.

Cloud-based testing services, such as AWS Device Farm, Google Firebase Test Lab, and BrowserStack, provide access to a vast array of real devices for testing. These services allow developers to run their apps on many devices and platforms without owning or maintaining the physical hardware.

Automation is essential when testing on different devices and platforms. Automated testing frameworks, like Appium, Espresso, and XCUITest, can execute tests across various devices and platforms, saving time and ensuring consistent test coverage.

User interface (or UI) as well as user experience (or UX) testing are vital for device and platform testing. UI elements should adapt to various screen sizes and resolutions, and the app's overall design should provide a consistent and intuitive experience across different devices.

For apps intended for global audiences, localization and internationalization testing are critical. It ensures the app functions correctly in different languages and cultures, accommodating varying text lengths and cultural expectations.

Continuous integration and continuous testing (CI/CT) practices are essential for maintaining quality and compatibility across devices and platforms. CI/CT pipelines automate the testing process, ensuring that tests are consistently executed whenever code changes are made.

User feedback and real-world testing are invaluable for uncovering issues that may not surface in controlled testing environments. Beta testing and soliciting user feedback on diverse devices and platforms can reveal compatibility problems and usability concerns.

Testing on different devices and platforms is a challenging yet imperative aspect of mobile app development. It ensures that apps function seamlessly and consistently across the diverse landscape of mobile devices and operating systems. In an era where user satisfaction and app quality are paramount, developers must embrace comprehensive testing practices to deliver apps that excel in compatibility and quality assurance. By doing so, they can confidently navigate the complexities of the mobile ecosystem and provide users with a superior and consistent app experience, regardless of their chosen device or platform.

CHAPTER VIII

Advanced Security Measures

Understanding mobile app security threats

Our daily lives are now dramatically impacted by mobile apps, serving many purposes, from communication and productivity to entertainment and finance. This ubiquity, however, has made mobile apps a prime target for malicious actors aiming to exploit vulnerabilities and compromise user data. Understanding mobile app security threats is crucial for developers, organizations, and users alike, as it enables them to safeguard the mobile ecosystem. In this section, we will explore the landscape of mobile app security threats, the risks they pose, and strategies to mitigate them, ensuring mobile app users' continued trust and safety.

Mobile applications have revolutionized our interaction with technology, but this revolution has also brought complexity. Users today access apps on a wide range of devices, including smartphones, tablets, wearables, and more. These devices run on different operating systems, such as iOS, Android, and others, each with multiple versions and configurations. To provide a seamless user experience and reach a broad audience, apps must function correctly on this diverse array of devices and platforms.

Mobile app security threats encompass many risks and vulnerabilities, including data leakage, malware and spyware, code vulnerabilities, man-in-the-middle (MitM) attacks, phishing attacks, unsecure APIs, and jailbreaking/rooting techniques. Each of these threats poses unique challenges to the security and integrity of mobile apps.

Mobile app security breaches can have severe consequences for both users and organizations. Users may suffer financial losses, identity theft, and privacy invasion, eroding their trust in app providers. For organizations, data breaches can result in financial penalties, legal liabilities, damage to reputation, and loss of customer trust.

The foundation of mobile app security lies in a proactive approach, where security is integrated into the app development process from the outset. This involves secure coding practices, threat modeling, risk assessment, and regular security audits. By considering security at every stage of development, developers can reduce the likelihood of vulnerabilities slipping through the cracks.

It is crucial to make sure that an app's features and data are only accessible to authorized users. Implementing robust authentication mechanisms, such as multi-factor authentication (MFA), and proper authorization controls can thwart many security threats. By verifying the identity of users and limiting their access to only what is necessary, apps can protect sensitive information from unauthorized access.

To prevent unwanted access, sensitive data should be encrypted whether it's in transit and at rest. Strong encryption algorithms and key management practices are essential components of this defense. Encryption provides an additional layer of security, ensuring that even if data is intercepted, it remains unintelligible to attackers.

App developers must handle data storage carefully, avoiding insecure storage locations and encrypting stored data. This prevents unauthorized access to sensitive user information. By protecting data at rest, apps can maintain the confidentiality of user data even if the device is compromised.

Secure APIs are critical for mobile apps that rely on external services. Developers should use proper authentication and authorization mechanisms, implement rate limiting, and monitor API usage for anomalies. By ensuring the security of API connections, apps can protect against data breaches and unwanted access to sensitive data.

Maintaining an app's security requires continuous vigilance. Regularly updating the app to fix security vulnerabilities and applying security patches to the underlying operating system and libraries is essential. Keeping all components up to date reduces the attack surface and helps protect against known vulnerabilities. Apps should request only the permissions necessary for their core functionality. Users should be informed about the data the app collects and given the option to grant or deny permissions. This transparency empowers users to make knowledgeable decisions about their privacy and security.

Conducting penetration testing as well as vulnerability scanning can help identify and remediate security weaknesses in an app. Ethical hackers and security professionals can simulate attacks to uncover vulnerabilities before malicious actors do. Regular testing and assessment are essential for maintaining robust security.

User awareness is a critical component of mobile app security. Educating users about common threats, safe browsing practices, and the importance of updating their apps and devices can help mitigate risks. Users who are informed and vigilant are less likely to be a victim of social engineering attacks or phishing attempts.

Mobile app security threats pose a significant challenge in an increasingly digital and interconnected world. Developers, organizations, and users all play vital roles in ensuring the security of the mobile ecosystem. By adopting a security-by-design approach, implementing best practices, and staying vigilant, we can safeguard the trust and safety of mobile app users. Mobile apps have enriched our lives in numerous ways, and securing them is not just a responsibility but a necessity to ensure continued innovation and progress in the mobile app landscape.

Implementing advanced security measures

Mobile app development is a rapidly evolving landscape driven by innovation and user experience. However, security should never be an afterthought amid this whirlwind of creativity. Mobile apps handle an abundance of sensitive information, from personal data to financial details, and ensuring the security and privacy of this data is paramount. This section delves into the world of advanced security measures in mobile app development, exploring the challenges developers face, the evolving threat landscape, and strategies to implement robust security measures to safeguard user data and trust.

The threats that mobile apps face are always changing along with the digital ecosystem. Due to the extensive utilization of mobile devices and the growing expertise of cybercriminals, mobile applications are frequently the target of security risks. These threats include malware, data breaches, identity theft, and unauthorized access, all of which can have severe consequences for both users and organizations.

Mobile app development presents unique security challenges. Developers must contend with diverse platforms and devices, from iOS and Android to devices with varying screen sizes and operating system versions. Protecting sensitive user data, implementing secure authentication and authorization mechanisms, securing external APIs, and educating users about security best practices are among the many considerations developers must address.

Advanced security measures are essential to tackle these challenges and protect user data. Secure coding practices that minimize vulnerabilities, encryption of sensitive data (both in transit and at rest), and implementing of biometric authentication methods such as fingerprint or face recognition are key components of a robust security strategy. Additionally, tokenization can reduce the risk of data exposure, while multi-factor authentication (MFA) adds an extra layer of security.

Security headers, such as Content Security Policy (CSP) and HTTP Strict Transport Security (HSTS), can protect against common web-based attacks like cross-site scripting (XSS) and man-in-the-middle (MitM) attacks. Regular security audits and penetration testing are vital for identifying and addressing vulnerabilities. User education is also crucial in reducing the risk of being a victim to social engineering attacks.

Adherence to data protection regulations, such as the CCPA and GDPR, is still another essential component of mobile app security. Compliance ensures that user data is handled per legal requirements, helping build trust with users.

Building trust through security measures is essential. Mobile apps with a strong reputation for security and privacy are more likely to attract and retain users. This involves transparent privacy policies, secure data handling, prompt response to security incidents, and a commitment to ongoing security updates.

In conclusion, mobile app security is not a mere checkbox on a development checklist but a fundamental requirement. Implementing advanced security measures is about more than just compliance; it is about building a resilient defense against an ever-evolving threat landscape. By prioritizing security from the earliest stages of app development and remaining vigilant in the face of new challenges, developers can create mobile apps that offer exceptional user experiences while ensuring that their data is safe and their privacy is respected. Trust and security are inextricably linked in the mobile app ecosystem, and both are essential for long-term success.

Encryption and data protection

In today's digital age, mobile apps have transformed how we interact with information and services, making our smartphones indispensable tools for communication, finance, healthcare, and more. However, this convenience comes with a trade-off: the vast amount of sensitive data handled by these apps. Users entrust their most private information to mobile apps, from personal information to financial details and even biometric data. Mobile app developers must prioritize data protection to fulfill this trust and meet ethical and legal obligations. Encryption, a fundamental tool in the realm of data security, plays a central role in safeguarding user privacy and securing sensitive information in mobile apps.

Encryption, at its core, transforms readable data, or plaintext, into an unreadable format known as ciphertext. This transformation is achieved through mathematical algorithms and cryptographic keys. The ciphertext can only be deciphered back into plaintext by someone possessing the corresponding decryption key. The strength of encryption lies in the complexity of these algorithms and the secrecy of the encryption keys.
In mobile app development, encryption serves two primary purposes: protecting data at rest and data in transit. Data at rest refers to information stored on the device or remote servers, while data in transit relates to information transmitted between the app and external servers or services.

When data is at rest, it is vulnerable to unauthorized access if not adequately protected. Encryption ensures that this data is stored in an encrypted format. Even if attackers can access the storage medium, they cannot decipher or utilize the data without the decryption key. This level of protection prevents unauthorized access to stored data.

Similarly, during data transmission between the app and external servers (such as during login or financial transactions), the data is vulnerable to interception by eavesdroppers. Encryption guarantees that this data is transmitted in an encrypted format, rendering it indecipherable to anyone lacking the decryption key. This protective measure thwarts interception and tampering attempts, securing the transmitted data's integrity and confidentiality.

Two primary types of encryption are commonly used in mobile app development: symmetric and asymmetric encryption. Symmetric encryption is effective since it uses the same key for decryption and encryption, but it does necessitate a safe way for the sender and recipient to share the key. On the other hand, asymmetric encryption uses a pair of keys—a public key for encryption and a private key for decryption. While this approach eliminates the need for securely sharing a single key, it involves more computational overhead.

Implementing encryption in mobile apps requires adherence to best practices. Key management is essential; only authorized processes or individuals should have access to the keys, which must be kept safely. The choice of encryption algorithms is equally crucial, with developers advised to opt for well-established and secure algorithms, steering clear of deprecated or weak options. Data classification helps determine which data should be encrypted and to what extent, recognizing that not all data may require the same level of protection. Secure communication protocols like HTTPS should be employed for data transmission to ensure end-to-end encryption, and regular updates are essential to address encryption vulnerabilities and weaknesses that may emerge over time.

Apart from adhering to encryption best practices, developers of mobile applications also need to take data protection regulations into account, which include CCPA, HIPAA, and GDPR. Compliance guarantees that user data is handled in accordance with legal requirements and industry standards, further bolstering data protection efforts.

In conclusion, encryption is not merely a security measure in mobile app development; it is the foundation of trust between users and app providers. It instills confidence in users, assuring them that their personal and sensitive information is safe from prying eyes and unauthorized access. For developers and organizations, encryption is a commitment to ethical data handling and responsible stewardship of user information. In an ever-evolving landscape of mobile app development, encryption remains a beacon of security and a shield against data breaches, ensuring that mobile apps remain trusted guardians of user privacy and security.

Security best practices for APIs and user authentication

In the realm of mobile app development, ensuring the security of user data is paramount. This task is further complicated by the reliance on external services through Application Programming Interfaces (APIs) and the need for robust user authentication mechanisms. This section will explore the security challenges associated with APIs and user authentication in mobile app development and discuss best practices to safeguard user data and maintain user trust.

APIs serve as the bridges connecting mobile apps with external servers, services, and databases. While they enable seamless experiences, they also pose potential security risks. To mitigate these risks, robust API security is crucial. Data breaches, unwanted access, and other security issues can be prevented through rigorous authentication mechanisms.

User authentication is the first line of defense in API security. Mobile apps should implement strong authentication methods, including multi-factor authentication (MFA), whenever possible. Utilizing strong passwords, biometric authentication (such as fingerprint or facial recognition), and token-based authentication ensures that only authorized individual can access the app and its associated APIs.

Securing the communication between mobile apps and APIs is vital. Data transmission must be encrypted using Transport Layer Security (TLS) to prevent eavesdropping and data interception, especially when transmitting sensitive details like login credentials or payment data.

API key management is another critical aspect. API keys, which authenticate apps to the server, should be stored securely. Exposed or leaked API keys can be exploited by attackers, leading to unauthorized access and potential data breaches.

Rate limiting and throttling mechanisms should be implemented to prevent abuse of APIs. These controls restrict the number of API requests allowed within a specific time frame, protecting the server from overload and reducing the risk of malicious attacks.

Input validation and data sanitization are essential to prevent injection attacks (e.g., SQL injection) and protect against cross-site scripting (XSS) vulnerabilities. Mobile apps must validate and sanitize user inputs before sending them to APIs.

Access controls and authorization mechanisms should be stringent, ensuring that users can only access data and functionality for which they are authorized. Role-based access control (RBAC) is a common approach to managing permissions.

Effective user authentication relies on several best practices. Strong password policies, including complex passwords and regular password changes, enhance security. Biometric authentication methods, including fingerprint or facial recognition provide added security and user convenience.

OAuth 2.0 and OpenID Connect are recommended for secure authentication and authorization, especially when integrating with third-party services like social media logins. Token-based authentication, such as JSON Web Tokens or OAuth tokens, ensures secure management of user sessions and authorization.

Secure session management practices protect user sessions from hijacking or session fixation attacks. These practices help maintain the integrity of user authentication.

API security presents unique challenges in mobile app development, especially when integrating with third-party APIs over which developers have limited control. Securely storing keys, tokens, and secrets on mobile devices is a challenge, as attackers may attempt to extract these credentials.

Protecting user privacy is another critical consideration. Developers must ensure that user data is handled in compliance with data protection regulations, respecting user rights and expectations.

Monitoring and logging API activities are essential to promptly detect and respond to suspicious behavior or security incidents. Timely identification and response can mitigate potential security threats.

In conclusion, security is not an optional feature but an ongoing commitment in mobile app development. Robust API security and user authentication are integral components of this commitment. By implementing best API security and user authentication practices, mobile app developers can protect sensitive user data, prevent security breaches, and maintain user trust. In a landscape where data breaches and security incidents are prevalent, these practices are essential for mobile apps' long-term success and reputation. As the mobile app ecosystem evolves, security must remain a top priority, ensuring that users can engage with apps confidently, knowing their data is secure and their privacy is respected.

CHAPTER IX

Advanced User Engagement and Analytics

Strategies for user engagement and retention

Today, mobile apps are a necessary component of our everyday existence, providing us with services, entertainment, and solutions to various needs. However, the success of a mobile app isn't solely determined by its functionality or aesthetics; it hinges on user engagement and retention. In this section, we will explore crucial strategies for keeping users engaged and retained in the fast-paced world of mobile app development.

User engagement is the interaction and involvement users have with an app. It encompasses regularly opening the app, completing tasks, and spending time within it. On the other hand, user retention is the ability to keep users returning to the app over an extended period. Engagement and retention are critical for a mobile app's long-term success and sustainability.

The foundation of user engagement and retention lies in user-centric design. Developers must prioritize creating an intuitive and user-friendly interface. Conducting user research, gathering feedback, and continuously iterating on the app's design based on user preferences and behavior are essential practices.

Personalization is a powerful tool for engaging users. Tailoring the app's content and features to match individual preferences increases user satisfaction and the likelihood of repeated use. Customization options, such as themes, settings, and notifications, give users a sense of control over their app experience.

Strategic utilization of push notifications and in-app messaging can re-engage users and keep them informed about new features, content, or special promotions. However, balancing valuable updates and spammy notifications is crucial, as excessive messaging can lead to user annoyance.

Gamification elements like points, badges, and leaderboards can make the app experience more engaging and competitive. These elements tap into users' intrinsic motivations and encourage them to return to the app to achieve goals or compete with others.

High-quality content is the lifeblood of many apps, such as news, social media, and entertainment platforms. Consistently delivering fresh and relevant content keeps users engaged. Developers should invest in creating and curating content that aligns with user interests and needs.

Integrating social features, such as sharing, commenting, and connecting with friends, can enhance user engagement. Social interactions within the app foster a sense of community and give users reasons to return and interact with others.

A slow and buggy app can quickly deter users. Optimizing app performance, ensuring fast loading times, and minimizing crashes are fundamental for retaining users. Users should find the app responsive and reliable in all circumstances.

Developers can test various features and designs using A/B testing to figure out which ones result in higher rates of engagement and retention. Analytics tools provide insights into user behavior, helping developers make data-driven decisions to improve the app's performance.

An effective onboarding process and user-friendly tutorials can help users understand the app's features and functionalities. A smooth introduction to the app reduces frustration and encourages users to explore further. Implementing loyalty programs, offering rewards, or providing exclusive access to loyal users can incentivize them to remain engaged and committed to the app. Rewarding users for their continued participation fosters a sense of belonging.

Encouraging user feedback and offering responsive customer support create a positive user experience. Users who feel heard and supported are more likely to stay engaged with the app, even if they encounter issues.

The mobile app landscape is dynamic, with new technologies and trends emerging regularly. Stagnation can lead to user attrition. Developers should continuously innovate, staying updated with the latest trends and technologies to keep the app fresh and competitive.

Mobile app developers face several challenges in their quest to engage and retain users. These include intense competition, short user attention spans, changing user preferences, and the need to balance user privacy with personalized experiences.

In the ever-evolving world of mobile app development, user engagement and retention are paramount. By adopting user-centric design, personalization, push notifications, gamification, and other strategies, developers can create an app that users download and return to regularly. Success in mobile app development is not just about acquiring users; it's about building and nurturing long-term relationships with them. In this pursuit, developers must listen to user feedback, adapt to changing preferences, and leverage technology to provide experiences that captivate and retain users in an increasingly competitive landscape. Ultimately, engaging and retaining users is the hallmark of a successful and enduring mobile app.

Advanced analytics tools and techniques

In the fast-evolving landscape of mobile app development, success hinges on creating aesthetically pleasing and functional apps and understanding and engaging users effectively. Advanced analytics tools and techniques have emerged as essential components of mobile app development, offering developers valuable insights into user preferences, behavior, and app performance. In this section, we will delve into the world of advanced analytics in mobile app development, exploring the significance of data-driven decision-making, the tools available, and the techniques used to extract actionable insights for optimizing user experiences and app performance.

Data-driven decision-making has revolutionized the way mobile apps are developed and maintained. It involves gathering and analyzing data to guide development choices, from feature enhancements to user engagement strategies. By harnessing the power of advanced analytics, developers can gain an extensive understanding of user behavior, identify pain points, and make informed decisions that lead to improved user experiences and app performance.

User analytics tools provide insights into how users interact with an app. Metrics like user demographics, app usage patterns, session duration, and retention rates help developers understand their users and how they engage with the app. Advanced analytics tools offer segmentation capabilities, allowing developers to group users based on particular criteria, such as location or behavior, to target them with tailored experiences and features.

Funnel analysis tracks user journeys within the app, mapping users' steps from initial interaction to desired actions, such as making a purchase or completing a registration. By identifying drop-off points in the funnel, developers can pinpoint areas where user experiences need improvement and implement changes to increase conversion rates.

Heatmaps visually represent user interactions within an app, highlighting high and low engagement areas. Session recording tools capture real user interactions, providing developers with a granular view of how users navigate the app. Heatmaps and session recordings help identify usability issues that need attention, such as confusing UI elements or navigation bottlenecks.

Crash reporting and error tracking tools are critical for promptly identifying and promptly addressing app crashes and bugs. They provide insights into the root causes of crashes, enabling developers to prioritize and fix issues that impact user experience and app stability.

A/B testing and experimentation tools allow developers to conduct controlled experiments by presenting different app versions to distinct user groups. This technique helps assess the impact of changes to features, designs, or user flows on key metrics like user engagement, retention, and conversion rates. By comparing the performance of different variants, developers can make data-driven judgments about which changes to implement permanently.

Predictive analytics leverages historical data to predict future user behavior and trends. Machine learning models can determine patterns and predict user preferences or potential churn. By anticipating user needs and preferences, developers can proactively tailor the app experience to meet those expectations.

Monitoring app performance is essential for ensuring smooth user experiences. Tools for app performance monitoring track metrics like load times, latency, and resource consumption. Identifying and optimizing performance bottlenecks is crucial for preventing user frustration and abandonment.

As app developers gather and analyze user data, it's imperative to prioritize data privacy and comply with regulations like GDPR and CCPA. Advanced analytics tools should include features for data anonymization, user consent management, and data access controls to safeguard user information and ensure legal compliance.

Real-time analytics provide developers with instant insights into user interactions and app performance. This capability is particularly valuable for detecting and responding to issues as they happen, ensuring a seamless user experience.

Despite the benefits, advanced analytics in mobile app development come with challenges. Gathering and interpreting vast amounts of data can be overwhelming, leading to information overload. Additionally, ensuring the accuracy and security of user data is a paramount concern. Developers must also navigate the complexity of integrating multiple analytics tools into the app.

In the dynamic world of mobile app development, advanced analytics tools and techniques have transformed how developers understand and engage users. By harnessing the power of user analytics, funnel analysis, heatmaps, session recording, crash reporting, A/B testing, predictive analytics, and app performance monitoring, developers can make data-driven decisions that enhance user experiences and drive app success.

However, with great analytical power comes great responsibility. Developers must prioritize data privacy and compliance while navigating the complexities of integrating and managing multiple analytics tools.

As the mobile app ecosystem evolves, advanced analytics will remain an indispensable part of the development process. Embracing these tools and techniques empowers developers to create apps that meet user needs and iterate and improve continuously, ensuring that their apps thrive in a competitive and ever-changing marketplace. Ultimately, advanced analytics are the compass that guides developers toward delivering mobile apps that captivate users and drive long-term success.

Implementing push notifications and in-app messaging

In the dynamic world of mobile app development, staying connected with users and delivering timely and relevant information is essential for success. Push notifications and in-app messaging have emerged as indispensable tools to achieve this goal. These communication channels are direct links between mobile apps and their users, facilitating real-time updates, reminders, promotions, and personalized content that foster user engagement and retention. Their implementation drives app success by re-engaging users, boosting app usage, and enhancing the overall user experience.

Push notifications are messages delivered from a mobile app to a user's device, even when the app is not actively in use. They appear as notifications on the device's lock screen, home screen, or notification center, grabbing the user's attention. Push notifications can serve various purposes, including delivering information updates, promotions and offers, encouraging user engagement, and re-engaging inactive users. They are a powerful tool to keep users informed and engaged.

On the other hand, in-app messaging involves sending messages or prompts to users while they are actively using the app. These messages appear within the app's user interface and can take various forms, such as onboarding tutorials, personalized recommendations, feedback and surveys, and announcements and updates. In-app messaging guides users through the app, provides relevant content, collects valuable feedback, and keeps users informed about app changes and enhancements.

Successful implementation of push notifications and in-app messaging requires a thoughtful approach. Personalization is critical to delivering messages that resonate with users. Leveraging user data and segmentation enables developers to send tailored notifications and in-app messages based on user behavior, preferences, location, and demographics. This personalization enhances relevance and engagement.

Timing and frequency are critical considerations. Sending notifications at the right moment increases the likelihood of user engagement. Developers should avoid overwhelming users with excessive notifications and implement frequency capping to limit the number of messages sent in a given timeframe. A well-timed message can make a significant difference in user response.

A/B testing is a beneficial strategy for optimizing message content, formats, and timing. Experimenting with different variations allows developers to gather data-driven insights and refine their messaging strategies. It helps determine what resonates best with the target audience and enhances the effectiveness of notifications and in-app messages.

Respecting user privacy and preferences is essential. Developers should provide clear opt-in options for push notifications and obtain user consent before sending messages. It is equally important to make it easy for users to opt out at any time. Striking a balance between customization and intrusiveness is crucial in building user trust.

Deep linking is another useful technique in message implementation. It allows developers to direct users to specific app screens or content when they interact with a notification. This reduces friction and provides users with direct access to relevant information, enhancing the overall user experience.

The impact of push notifications and in-app messaging on user engagement and retention is significant. Well-timed and relevant messages boost user engagement by encouraging app opens and interactions. Regular communication through notifications and in-app messages helps retain users by reminding them of the app's value and encouraging return visits. Personalized messages and guidance also improve the overall user experience, making users feel valued and understood. Targeted promotions and recommendations can increase conversion rates, revenue, and user satisfaction.

However, these communication tools come with challenges. Overusing them or sending irrelevant messages can annoy users and lead to uninstallations or opt-outs. User privacy concerns are rising, and developers must handle user data responsibly and transparently to maintain trust. Additionally, user preferences regarding message frequency and content can vary, requiring developers to balance customization and intrusiveness.

In conclusion, push notifications and in-app messaging have transformed how mobile apps engage and retain users. When implemented strategically, these tools strengthen user connections, foster engagement, and boost retention rates. Developers must approach their use carefully, ensuring personalization, timing, and relevance to provide value without overwhelming users. In an ever-evolving mobile app landscape, push notifications and in-app messaging remain indispensable for building stronger, more enduring relationships between apps and their users.

A/B testing and user feedback integration

In the ever-evolving world of mobile app development, the quest for success centers on creating apps that resonate with users, providing seamless experiences, and meeting their needs effectively. Two powerful strategies have emerged to guide developers toward these goals:

A/B testing and user feedback integration. These techniques are pivotal in the iterative process of app development, as they empower developers to optimize user experiences, enhance app performance, and ultimately drive app success.

A/B testing, often called split testing, is a systematic approach to comparing multiple versions of an app or specific features to determine which variant performs better in terms of user engagement, retention, conversion rates, or other critical metrics. This process entails randomly assigning users to different groups, each exposed to a distinct app or feature variant. By meticulously analyzing the performance data collected from each group, developers can make data-driven decisions about which variant should be permanently implemented.

A/B testing is not merely about identifying the better option; it is a crucial tool for optimizing user experiences. Developers can experiment with various design elements, layouts, colors, and content to fine-tune the app's user interface and overall design. For instance, they can compare app layouts to ascertain which drives higher user engagement or conversion rates. This iterative approach ensures that user preferences and behaviors are considered throughout the app's development lifecycle.

Additionally, A/B testing is instrumental in assessing the effectiveness of specific features within an app. Developers can gauge how well individual features resonate with users by comparing user interactions and outcomes between different feature variants. This insight empowers developers to prioritize feature development efforts effectively, allocating resources to the most impactful features, contributing to user satisfaction and engagement.

Furthermore, A/B testing is pivotal in optimizing app conversion funnels, such as user registration, subscription sign-ups, or in-app purchases. By experimenting with variations of these funnels, developers can pinpoint the design or flow that encourages users to complete desired actions. This optimization results in improved conversion rates, subsequently leading to increased revenue and user engagement.

User feedback integration is equally vital in mobile app development, as it involves actively collecting and incorporating user input, suggestions, and complaints into the app development process. This approach is rooted in creating channels for users to provide feedback, such as in-app surveys, feedback forms, or direct contact with support teams. By leveraging user feedback, developers gain invaluable insights into user preferences, pain points, and areas that require improvement.

One of the primary benefits of user feedback integration is its ability to identify pain points within the app. Users often encounter usability issues, bugs, or areas where they get stuck. Developers can quickly identify these pain points by actively gathering feedback and prioritizing fixes and enhancements. This, in turn, results in a smoother and more enjoyable user experience, fostering user satisfaction and loyalty.

Additionally, user feedback integration enhances user satisfaction by demonstrating a commitment to addressing user concerns and suggestions. Users are more likely to remain associated with the app and become devoted when they feel that their opinions are heard and taken into consideration. This sense of partnership between users and developers can significantly impact user retention and overall app success.

User feedback integration promotes a culture of continuous improvement in app development. Developers can use feedback to inform ongoing iterations and updates, ensuring the app remains aligned with evolving user needs and preferences. It acts as a dynamic feedback loop, enabling developers to adapt swiftly to changing user expectations and market dynamics.

The integration of A/B testing and user feedback is a powerful approach to app development. These two strategies complement each other, offering a comprehensive view of user preferences and behavior. A/B testing provides quantitative insights into user interactions and behavior, while user feedback offers qualitative insights into user sentiments, suggestions, and concerns.

Developers can validate findings more effectively by combining A/B testing and user feedback integration. For example, user feedback can serve as a means to validate or explain the results of A/B tests. If an A/B test indicates that a particular feature variant leads to higher engagement, user feedback can provide additional context and insights into why users prefer that variant.

Moreover, integrating A/B testing and user feedback enables developers to iterate and improve with precision. A/B testing identifies areas for improvement, and user feedback provides specific guidance on how to make those improvements. Developers can use user feedback to refine design elements, enhance features, and improve usability, resulting in an app that truly resonates with users.

The integration also facilitates the prioritization of development efforts. When users consistently request a specific feature or improvement, developers can prioritize testing variations related to that feature to determine the most effective implementation. This approach ensures that development efforts are aligned with user demands and preferences.

Furthermore, integrating A/B testing and user feedback enhances the user-centric nature of app development. It fosters a culture where data-driven decisions are augmented by direct user insights, resulting in apps that function well and meet and exceed user expectations. This user-centricity is a hallmark of successful and sustainable app development.

However, integrating A/B testing and user feedback is not without its challenges. It requires careful planning to align testing and feedback collection with development timelines. Developers must ensure that user feedback is collected systematically, analyzed effectively, and translated into actionable improvements. Maintaining a balance between quantitative data from A/B tests and qualitative insights from user feedback can be demanding, but it ultimately leads to more informed and effective decisions.

In conclusion, A/B testing and user feedback integration are integral to mobile app development. These strategies empower developers to optimize user experiences, enhance app performance, and drive app success. By continuously iterating based on A/B test results and user feedback, developers create apps that meet user needs and exceed their expectations. These tools provide a roadmap to app excellence and user satisfaction in an ever-evolving mobile app landscape.

CHAPTER X

Monetization and App Distribution

Advanced monetization strategies

One essential component of developing a mobile app is monetization, as it determines the sustainability and profitability of an app. In the fiercely competitive app market, developers must go beyond traditional monetization methods to stay ahead. Advanced monetization strategies have emerged as essential tools for maximizing revenue while providing value to users. This section will explore the significance of advanced monetization strategies in mobile app development, examine various approaches, and discuss how developers can balance profitability and user satisfaction.

Monetization strategies in mobile app development have come a long way since the early days of paid apps and in-app advertising. As the app ecosystem has matured, developers have recognized the need for more sophisticated approaches that generate revenue and enhance the user experience.

In-app purchases have revolutionized mobile app monetization by allowing users to buy virtual goods, premium features, or subscriptions within the app. This strategy capitalizes on the "freemium" model, where the app is free to download and use, but users can enhance their experience by making purchases. IAPs are particularly effective in gaming apps, where users can buy items, power-ups, or skins to enhance gameplay.

However, IAPs are not limited to games and are increasingly used in various app categories, including productivity, fitness, and entertainment.

Subscription-based monetization models have gained prominence, especially for apps offering premium content or services. Subscriptions provide a steady revenue stream, allowing developers to improve and update their apps continuously. Popular subscription-based apps include streaming services like Netflix and Spotify, which offer access to exclusive content and features for a monthly fee. This model is effective for retaining users and building a loyal customer base.

While traditional banner ads are still prevalent, advanced in-app advertising strategies have emerged. These include native ads that blend seamlessly with the app's content, rewarded video ads that offer users incentives for watching, and interstitial ads that appear at natural breaks in the user experience. Advanced ad networks and programmatic advertising have made it easier for developers to optimize ad placement and maximize revenue without compromising the user experience.

Collaborating with brands and sponsors can be a lucrative monetization strategy. Developers can integrate sponsored content or promotions into their apps, aligning with the app's theme or target audience. These partnerships can take various forms, including sponsored challenges in fitness apps or branded filters in photo editing apps. Successful sponsorships generate revenue and add value to the user experience when executed tastefully.

Data monetization involves leveraging user data, with their consent, to provide personalized experiences or insights to third-party companies. For instance, weather apps can provide location-based data to local businesses for targeted advertising. Although data monetization can be profitable, developers need to prioritize user privacy and abide to data protection regulations such as the CCPA and GDPR.

Many apps introduce virtual currencies or gift card systems, allowing users to purchase and redeem digital credits for various purposes. These virtual currencies can unlock premium content, send virtual gifts to other users, or participate in virtual economies within the app. This strategy encourages user engagement and spending while enhancing the app's financial viability.

While advanced monetization strategies offer numerous opportunities to increase revenue, developers must navigate the fine line between profitability and user experience. Overzealous monetization efforts can lead to user frustration, app abandonment, or negative reviews. Therefore, it is crucial to implement these strategies thoughtfully and ethically.

User experience should remain at the forefront of app development. Monetization strategies should enhance, rather than hinder, the user experience. Intrusive ads, aggressive upselling, or paywalls that limit essential features can lead to user dissatisfaction. Developers should prioritize creating value for users and ensuring that paid options provide genuine benefits.

Transparent communication with users is essential. Clearly explain how monetization methods work, what users can expect, and how their data will be used. Provide opt-in options for ads, data sharing, or subscription services. Users who willingly engage with monetization features are more likely to perceive the app positively.

A/B testing and user feedback play a crucial role in optimizing monetization strategies. Developers can experiment with different approaches and analyze user behavior to determine the most effective strategy. User feedback can provide insights into the impact of monetization on the overall user experience.

If an app offers premium features or content for purchase, pricing should be fair and aligned with the perceived value. Users should clearly understand the benefits of making a purchase. Developers should avoid creating a paywall that restricts core functionalities, which can alienate free users.

For apps that incorporate advertising, ensuring ad quality and relevance is crucial. Irrelevant or low-quality ads can disrupt the user experience and lead to ad fatigue. Developers should work with reputable ad networks and monitor ad content to maintain a high standard.

User privacy and data protection must be paramount when implementing data monetization strategies. Developers should obtain explicit user consent for data collection and sharing and comply with relevant data protection regulations. Transparent data practices build trust with users.

Advanced monetization strategies have transformed the mobile app development landscape, offering developers many opportunities to generate revenue while providing value to users. The evolution from paid apps to sophisticated models like in-app purchases, subscriptions, and data monetization reflects the industry's adaptability and commitment to sustainability. However, developers must carefully balance profitability and user experience to ensure their apps remain competitive, user-friendly, and ethically sound. By prioritizing user-centric design, transparency, fair pricing, and data protection, developers can thrive in the world of advanced monetization while maintaining user trust and satisfaction.

App store optimization (ASO) for increased visibility

In the crowded and highly competitive world of mobile app development, creating an exceptional app is just the first step. To succeed, developers must ensure their apps are discoverable and visible to their target audience. App Store Optimization (ASO) has emerged as a vital strategy to achieve this goal. ASO encompasses a range of techniques and best practices to improve an app's visibility in app stores, ultimately leading to increased downloads and user engagement. This section will delve into the significance of ASO in mobile app development, explore its key components, and discuss how developers can leverage ASO to boost their app's visibility and success.

Visibility is paramount in an app ecosystem where millions of apps are vying for users' attention. App stores, like Google Play as well as Apple App Store are the primary gateways for users to discover new apps. Therefore, an app's position and visibility within these stores greatly influence its chances of success. A well-optimized app will likely be discovered by potential users, leading to increased downloads, higher rankings, and improved user engagement.

ASO is a multifaceted strategy encompassing various components, each contributing to an app's visibility and attractiveness to users. These components include:

Keywords are the foundation of ASO. Developers must identify relevant keywords and phrases that potential users might use to search for their app. These keywords should be strategically placed in the app's title, subtitle (for iOS apps), and description. Researching and selecting the right keywords is an ongoing process that involves monitoring trends and user behavior.

The app's title and description are critical elements of ASO. A compelling and informative title that includes relevant keywords can significantly improve visibility. The description should concisely communicate the app's value proposition and key features while naturally incorporating relevant keywords.

Visual elements play a vital role in attracting users' attention. An eye-catching app icon and high-quality screenshots that showcase the app's functionality and user interface can influence users' decisions to download the app. A/B testing different icon and screenshot variations can help identify the most effective designs.

Positive user ratings and reviews enhance an app's credibility and impact its visibility. Encouraging satisfied users to leave reviews and ratings can boost an app's overall appeal. Developers should also address negative reviews promptly and use feedback to make improvements.

To reach a global audience, app localization is crucial. Translating the app's title, description, and keywords into multiple languages can broaden its reach and increase visibility among non-English-speaking users.
Frequent updates that introduce new features or improvements demonstrate a commitment to app quality and can improve rankings. Encouraging user engagement

through push notifications, in-app messages, or social media can increase user retention and more positive reviews.

Analyzing the ASO strategies of competitors can provide valuable insights. Understanding what keywords they target, how they position their apps, and which visual elements they use can help developers refine their own ASO efforts.

To harness the power of ASO and increase visibility in app stores, developers should adopt a systematic approach: Effective ASO starts with in-depth keyword research. Developers should identify relevant keywords that align with the app's purpose and functionality. Leveraging ASO tools and analytics can provide insights into search volume, competition, and trending keywords. Regularly update the list of targeted keywords to stay relevant.

The app's title and description should be optimized with carefully chosen keywords. The title should be concise and memorable, while the description should provide a clear value proposition. Keywords should be used naturally, avoiding keyword stuffing, which can lead to penalties by app stores.

Investing in an appealing app icon and high-quality screenshots is essential. Visual assets should accurately represent the app's features and benefits. A visually appealing app store listing is more likely to attract users' attention and encourage them to explore further.

Developers should actively encourage users to rate and review their app. This can be done through in-app prompts, follow-up emails, or incentives. Engaging with users who provide feedback, whether positive or negative, demonstrates responsiveness and a commitment to improvement.

ASO is an ongoing process that needs continuous monitoring and adjustment. Developers should regularly review app performance, track keyword rankings, and analyze user feedback. A/B testing different elements of the app store listing can help refine ASO strategies for better results.

For developers lacking ASO expertise, seeking professional ASO services or hiring an ASO specialist can be a worthwhile investment. These experts can provide valuable insights and implement effective strategies to enhance app visibility.

App Store Optimization (ASO) is a crucial component of mobile app development that directly impacts an app's visibility and success. In a competitive marketplace, ASO techniques empower developers to stand out, attract users, and increase downloads. By diligently optimizing keywords, fine-tuning app store listings, and fostering positive user interactions, developers can leverage ASO to ensure that their apps meet the needs of their target audience and reach them effectively. In the dynamic world of mobile app development, ASO is an indispensable tool for achieving sustainable growth and recognition.

Strategies for handling app updates

Mobile app development is a dynamic and ever-evolving field, and one of its essential aspects is the regular release of app updates. App updates are vital for keeping an app functional and secure and enhancing its features, improving user experience, and staying competitive in the market. However, handling app updates requires a well-thought-out strategy to ensure the process is smooth, efficient, and beneficial for developers and users.

The importance of app updates cannot be overstated. First and foremost, updates are critical for addressing bugs, glitches, and vulnerabilities that may emerge after an app's initial release. By promptly fixing these issues, developers ensure that their app remains stable and secure, which is essential for maintaining user trust.

App updates also serve as an opportunity to optimize performance. Over time, an app may become slower or less efficient due to changes in device hardware or operating system updates. Regular updates allow developers to fine-tune the app's performance, ensuring a smoother and faster user experience.

Moreover, updates bring about feature enhancements. Developers can use updates to add new features, improve existing ones, or introduce innovative elements that maintain consumers' interest in and excitement for the app. These enhancements not only satisfy current users but also attract new ones.

App updates are crucial for ensuring compatibility with evolving mobile operating systems. As operating systems advance, apps must adapt to these changes to remain compatible. Updates ensure that an app continues to work seamlessly on the latest OS versions, preventing compatibility issues that could lead to user frustration.

Regular updates also provide a competitive edge in the app market. They demonstrate a developer's commitment to improving their app, which can help the app stand out in a crowded market, attract new users, and retain existing ones.

To effectively handle app updates, several strategies can be employed. A regular update schedule is essential to build user trust and keep them engaged. Whether it's monthly, bi-monthly, or quarterly updates, consistency in updates fosters user trust and provides users with a sense of predictability.

User feedback is a valuable resource for identifying areas that need improvement or enhancement. Developers should actively encourage users to provide feedback and use it to inform their update priorities. Addressing user concerns and implementing requested features can boost user satisfaction and loyalty.

Versioning and changelogs are crucial for communicating what each update includes. Users should easily understand what changes, improvements, or additions they can expect. This transparency fosters trust and ensures users know the app's ongoing development.

Before a wider release, beta testing with a select group of users can help developers identify and resolve potential issues. This approach allows developers to release more polished updates and gives users a sense of involvement in the app's development.

A/B testing is a beneficial strategy for significant updates or changes to user interface elements. By comparing user behavior and feedback, developers can decide which design or feature variation is more effective.

Handling app updates requires careful consideration for different platforms. Each platform has its guidelines, release processes, and user demographics that may influence update strategies.

App Store Optimization (ASO) should not be overlooked when releasing updates. The app's metadata, such as the app description and screenshots, should reflect the changes to attract new users who may discover the app through app stores.

Rolling out updates in phases can be beneficial in some cases. This approach allows developers to monitor the impact of updates on a smaller group and address any unexpected issues before a broader release.

Backward compatibility is crucial to avoid excluding a portion of the user base. Developers should ensure that updates are compatible with older device models and operating systems.

Finally, user education and notifications are essential to inform users about the update's benefits and encourage them to update the app. Using push notifications, in-app messages, or email communication can convey the value of the update and its relevance to users.

In conclusion, app updates are a fundamental aspect of mobile app development that contributes to app functionality, user satisfaction, and overall success. Employing strategies that prioritize user feedback, maintain transparency, and consider different platforms and user demographics ensures that updates are well-received and beneficial for both the app and its users. Regular, well-planned updates keep an app competitive and create a positive user experience that fosters user loyalty and engagement. In the rapidly evolving landscape of mobile apps, handling updates effectively is key to long-term success.

Legal and compliance considerations

The world of mobile app development is a dynamic and lucrative field, but it is also one fraught with legal and compliance considerations that developers must navigate. As apps proliferate across platforms and devices, developers must be acutely aware of the complex laws, regulations, and ethical standards governing their work. In this section, we will delve into the various legal and compliance aspects that mobile app developers must address, emphasizing the critical role these considerations play in ensuring their creations' success and legality.

In the realm of mobile app development, intellectual property rights are of paramount importance. These rights encompass copyrights, trademarks, and patents, and developers must exercise caution to protect their own IP while respecting that of others.

Copyrights are automatically granted to original creative content such as code, graphics, music, and text, as soon as it's created and fixed in a tangible medium. Developers must carefully adhere to copyright laws, refraining from using copyrighted material without proper authorization or licensing.

Trademark infringement is another pitfall that app developers must steer clear of. Before selecting a name or logo for an app, it's crucial to conduct thorough trademark research to ensure that the chosen branding elements do not infringe on existing trademarks.

Additionally, patents may come into play when apps incorporate patented technologies or processes. Developers should be vigilant about identifying and respecting patent rights, conducting patent searches when necessary to avoid unauthorized use of patented technology.

The issue of privacy and data protection has grown increasingly prominent in recent years, and app developers must prioritize user privacy while adhering to data protection regulations.

Privacy policies are a critical component of mobile apps. Developers should craft transparent and easily accessible privacy policies that inform users about data collection, storage, and usage practices. Users should be required to consent to these policies before proceeding with app usage.

Data security is paramount, and apps should employ robust measures to safeguard user information. Encryption, secure authentication methods, and regular security audits are essential to protect sensitive data from breaches and unauthorized access.

For apps with users in the European Union, compliance with the General Data Protection Regulation is not optional—it's mandatory. Developers must obtain explicit user consent for data collection, provide data access and deletion mechanisms, and promptly report any data breaches.

Ensuring that mobile apps are accessible to all users, as well as those with disabilities, is a legal requirement and an ethical imperative. Failure to make apps accessible can result in discrimination and legal consequences.

The Web Content Accessibility Guidelines (WCAG) set international standards for web accessibility, and app developers should strive to adhere to these guidelines to ensure that their apps are usable by individuals with disabilities. Moreover, in the United States, the Americans with Disabilities Act (ADA) mandates that public accommodations, including digital services like mobile apps, must be accessible to individuals with disabilities. Non-compliance can lead to legal action.

Mobile apps are global products, and developers must be mindful of international laws and regulations that may apply to their apps. Export controls, in particular, restrict the international distribution of specific technologies, especially in cases where apps include encryption or other sensitive components. Additionally, apps may need to comply with specific local laws and regulations in the countries where they are made available, covering content restrictions, data protection, and taxation.

Protecting consumers from deceptive practices and fraudulent apps is a responsibility that app developers must take seriously. At all costs, developers should avoid false advertising, ensuring that app descriptions, functionality, and features accurately represent the app's capabilities. Additionally, transparency in in-app purchases and microtransactions is crucial. Developers should communicate the costs and terms associated with these transactions to users and have clear refund and cancellation policies.

Specific industries, such as healthcare and finance, may be subject to additional regulatory requirements that app developers must adhere to. For instance, apps dealing with healthcare information must adhere with HIPAA (Health Insurance Portability and Accountability Act) in the United States, which mandates strict safeguards for protected health information. Similarly, banking, investment, or payment processing apps must adhere to financial regulations like Know Your Customer (KYC) and Anti-Money Laundering (AML) requirements. Apps directed toward children must comply with the Children's Online Privacy Protection Act (COPPA), which governs the collection of personal information from children under 13 in the United States.

The app stores, like Apple App Store as well as Google Play Store, have their own guidelines and policies that developers must follow. These guidelines encompass review criteria, content restrictions, and monetization policies. Non-compliance with these app store guidelines can result in the removal or limitation of an app.

Developers can mitigate legal risks by including comprehensive terms of use, end-user license agreements (EULAs), and app disclaimers. Terms of use should establish the rules governing the relationship between the user and the app, outlining user obligations, limitations of liability, and mechanisms for resolving disputes. EULAs are necessary for apps requiring installation or specific licensing terms, detailing how users can use the app and any associated limitations. Disclaimers can clarify the app's purpose, limitations, and potential risks associated with its use.

Given the complexity of legal and compliance considerations, many app developers seek legal counsel to navigate these issues effectively. Legal counsel with technology and intellectual property law expertise can provide invaluable guidance. Regular compliance audits can help identify potential legal issues and ensure that apps remain in accordance with relevant laws and regulations.

In conclusion, mobile app development is a multifaceted endeavor that extends far beyond coding and design. Developers must consider an array of legal and compliance considerations to ensure their apps' legality, integrity, and ethical soundness. By staying informed about the legal landscape, seeking legal guidance when necessary, and integrating responsible development practices, app developers can create products that thrive in the competitive marketplace and adhere to the highest legal and ethical standards. Responsible development is a legal imperative and a cornerstone of long-term success and user trust in the ever-evolving world of mobile app development.

CONCLUSION

Recap of key takeaways from the e-book

As we conclude this e-book on "Unleashing Mobile App Innovation: Mastering Mobile App Development-Advanced Techniques and Best Practices," it's essential to recap the key takeaways and insights discussed throughout the chapters. This e-book has provided a comprehensive guide to advanced techniques and best practices in mobile app development, covering various aspects of app architecture, user experience, development environments, and more. Let's revisit the main points and lessons learned from exploring this dynamic and ever-evolving field.

Throughout this e-book, we've emphasized the significance of advanced app architecture patterns. Understanding models like MVVM and MVP can lead to more scalable and maintainable apps. It's essential to separate concerns within your app's structure, choose the right architecture pattern, and employ reactive programming for efficient data flow management.

Developers must prioritize scalability and maintainability. Clean code, modular architecture, and effective code organization are crucial to ensure your app remains manageable as it grows. We've also explored real-world case studies of successful apps demonstrating well-designed architectures' benefits.

User experience is an important factor in app development. User-centered design, intuitive navigation, and consistent branding are key principles to follow.

Ensuring your app is accessible and responsive across different devices and platforms is equally important.

Our discussion of advanced UI design patterns highlighted techniques such as adaptive layouts and dynamic theming. These patterns enable developers to create visually appealing and engaging user interfaces that cater to various user preferences and devices.

Responsive design and adaptive layouts are vital to providing a consistent user experience across different screen sizes and orientations. The adaptability of your app ensures that users can enjoy it on various devices seamlessly.

Effective development environments, integrated development environments (IDEs), and code version control systems are instrumental in streamlining the app development process. Adopting best practices for using these tools can significantly enhance your development workflow.

We've emphasized the importance of code version control systems, notably Git, for collaboration and code management. Implementing branching, pull requests, and continuous integration (CI) pipelines can elevate code quality and teamwork.

Advanced debugging and profiling techniques are essential for identifying and resolving issues efficiently. We explored methods like remote debugging, performance profiling, and memory analysis to ensure app stability and optimal performance.

Managing data in complex mobile applications requires a thoughtful approach. We examined various data storage solutions, offline data synchronization, and encryption methods to ensure data security and reliability.

Advanced data storage solutions, including NoSQL databases and cloud-based options, offer scalability and flexibility for apps dealing with large volumes of data.

Offline data synchronization strategies enable users to access and update data even when offline, ensuring data consistency across devices and network conditions.

Security is paramount in app development. We explored encryption methods, authentication protocols, and best practices for securing user data and communications within mobile apps.

Effective API integration is essential for accessing external data and services. We discussed advanced techniques for handling RESTful and GraphQL APIs, emphasizing the importance of efficient network requests.

Managing user authentication and authorization is critical to protect sensitive data and features. We explored authentication methods, OAuth, and JWT, emphasizing secure practices.

Robust error handling and retry mechanisms are crucial for maintaining app reliability. We covered strategies for gracefully handling errors, implementing exponential backoff, and ensuring a resilient app.

Profiling and optimizing app performance are ongoing tasks. We discussed techniques for identifying performance bottlenecks, optimizing code, and ensuring a smooth user experience.

Reducing app size directly impacts user adoption. To minimize app size, we explored methods such as code splitting, asset optimization, and app bundle configuration.

Effective memory management is crucial for app stability. We covered advanced memory management techniques, including memory leak detection and efficient memory allocation.

Multithreading and concurrency are essential for responsive apps. We discussed the use of background threads, thread synchronization, and concurrent programming patterns.

A robust testing strategy is integral to app quality. We explored unit, integration, and UI testing, emphasizing the importance of test automation and comprehensive test coverage.

Testing on diverse devices and platforms is necessary to reach a broad audience. We discussed strategies for testing on various devices, simulators, emulators, and cloud-based testing solutions.

Security threats must be understood and mitigated. We examined common mobile app security threats like data breaches, code vulnerabilities, and social engineering attacks.

Implementing advanced security measures, including two-factor authentication (2FA), encryption at rest and in transit, as well as security audits, is essential to protect against security threats.

User engagement and retention are critical for app success. We explored strategies such as push notifications, in-app messaging, and personalization to keep users engaged.

Advanced analytics tools and techniques provide valuable insights. We discussed advanced analytics to inform decision-making and improve app performance, including A/B testing, cohort analysis, and user behavior tracking. Effective push notifications and in-app messaging can boost user engagement. We explored best practices for creating timely, relevant, and non-intrusive notifications.

User feedback and A/B testing are essential for continuous improvement. We discussed integrating user feedback into development and conducting effective A/B tests to optimize user experiences and app features. This e-book has covered various advanced techniques and best practices in mobile app development. These insights provide a comprehensive guide to creating successful mobile apps, from architectural patterns to security measures, from user experience to performance optimization. By applying these principles and continuously adapting to the evolving landscape of mobile technology, developers can deliver high-quality, secure, and user-friendly apps that resonate with users and stand the test of time. In a rapidly changing field, staying informed and adopting best practices is the key to mobile app development success.

Encouragement for further exploration and learning

As we conclude this comprehensive e-book on advanced mobile app development techniques and best practices, it's essential to highlight this field's endless possibilities and ever-evolving nature. While we've covered many topics and provided valuable insights, it's just the beginning of your journey as a mobile app developer. This section encourages further exploration and learning, emphasizing the importance of staying curious, adaptable, and committed to your growth in this dynamic industry.

The world of technology, particularly mobile app development, is marked by continuous innovation and change. New programming languages, frameworks, and tools emerge regularly. To stay relevant and competitive, embrace the concept of lifelong learning. Dedicate time to explore emerging trends and technologies, attend conferences, participate in online courses, and read industry publications. Learning doesn't stop when you finish a project or master a skill; it's a lifelong journey.

Innovation often arises from experimentation and thinking outside the box. Feel free to experiment with new ideas and concepts in your app development projects. Try out unconventional solutions, explore novel user experiences, and push the boundaries of what's possible. Innovation can set your apps apart and lead to breakthroughs that benefit you and your users.

The tech community is a vast and interconnected ecosystem. Collaborate with fellow developers, designers, and professionals from diverse backgrounds. Engage in open-source projects, join developer communities, and attend meetups or hackathons. Networking expands your knowledge and opens doors to potential collaborations, mentorships, and career opportunities.

The importance of app security cannot be overstated. The threat landscape constantly evolves, with new vulnerabilities and attack vectors emerging regularly. Keep yourself informed about the most recent security threats and best practices for mitigating them. Stay updated on encryption techniques, authentication protocols, and secure coding standards. Regularly audit your apps for security vulnerabilities and implement security measures proactively.

User feedback is a goldmine of insights. Encourage users to provide feedback on your apps and genuinely listen to their suggestions and concerns. Use this feedback to iterate and improve your apps continuously. Prioritizing user satisfaction leads to better apps and builds a loyal user base.

While technical skills are crucial, understanding the business side of app development is equally important. Familiarize yourself with app monetization strategies, app store optimization (ASO), marketing techniques, and user acquisition strategies. A well-rounded understanding of technical and business aspects will help you create successful and sustainable apps.

As you gain experience and expertise, consider becoming a mentor to aspiring developers. Sharing your knowledge and insights with others contributes to the growth of the tech community and reinforces your understanding of the subject matter. Mentorship can be a rewarding way to give back to the community and make a positive impact. The mobile app development enviroment is constantly evolving with new technologies like the augmented reality (or AR), virtual reality (or VR), Internet of Things (or IoT), and 5G connectivity. These technologies offer exciting opportunities for innovation. Stay curious and explore how these emerging technologies can be integrated into your apps to provide unique and valuable experiences.

App development often involves solving complex problems. Practice problem-solving skills by taking on challenging projects and breaking them into manageable tasks. This enhances your technical abilities and fosters a problem-solving mindset that can be employed to various aspects of life.

Finally, stay inspired and passionate about what you do. The tech industry is fueled by individuals who are genuinely excited about the possibilities it offers. Follow industry thought leaders, read inspirational stories of successful developers, and remind yourself of your work's impact on people's lives. A strong sense of purpose and enthusiasm can be a driving force in your journey.

In summary, mobile app development is a dynamic and rewarding landscape that offers endless opportunities for growth and innovation. While this e-book has provided a solid foundation of advanced techniques and best practices, remember that your journey as a developer is an ongoing adventure. Embrace learning, experimentation, collaboration, and adaptation as you explore the ever-expanding horizons of mobile app development. With dedication and a passion for excellence, you can make a significant impact in this thriving industry and contribute to shaping the future of technology.

Final thoughts on the future of mobile app development

As we look to the future of mobile app development, it's clear that this dynamic and rapidly evolving field will continue to shape how we interact with technology and the world around us. The journey we've embarked on in this e-book, exploring advanced techniques and best practices, is just a glimpse into the limitless possibilities that lie ahead. In these final section, let's contemplate the trends and innovations poised to define the future of mobile app development.

One of the most exciting frontiers in mobile app development is the integration of AR and VR technologies. These immersive experiences can potentially revolutionize industries ranging from gaming and entertainment to education and healthcare. Imagine apps that seamlessly blend the physical and digital realm, allowing users to interact with virtual objects in real-time or enter immersive virtual environments. Developers who embrace AR and VR will be at the forefront of this transformative wave.

AI and ML are becoming increasingly integral to mobile app development. These technologies empower apps to make intelligent decisions, provide personalized experiences, and automate tasks. From chatbots offering real-time customer support to recommendation engines that tailor content to individual preferences, AI-driven apps enhance user engagement and efficiency. As AI and ML advance, their integration into app development will be ubiquitous.

The IoT ecosystem is expanding rapidly, with an ever-growing number of interconnected devices. Mobile apps will be pivotal in facilitating communication and control within this ecosystem. Imagine controlling your smart home, tracking your fitness, and monitoring your vehicle through a single app on your mobile device. Seamlessly integrating and managing IoT devices will be a valuable skill for future app developers.

The rollout of 5G networks promises lightning-fast speeds and low latency, opening up new opportunities for mobile app development. Apps that rely on real-time data streaming, such as augmented reality experiences and remote collaboration tools, will significantly benefit from 5G connectivity. Developers should prepare to harness the capabilities of this technology to deliver responsive and data-intensive apps.

Cross-platform development frameworks like Flutter and React Native are gaining traction, allowing developers to create apps that operate on multiple platforms with a single codebase. This trend will likely continue, enabling developers to reach wider audiences efficiently. Embracing cross-platform development can streamline the app development process and reduce time-to-market.

Security will remain a top priority with the increasing reliance on mobile apps for sensitive tasks. Future app developers must be well-versed in advanced security measures, including biometric authentication, end-to-end encryption, and blockchain-based solutions. Developers will play a critical role in safeguarding user data and privacy as cybersecurity threats evolve.

Voice-enabled apps and natural language processing are becoming more sophisticated. Virtual assistants like Siri, Google Assistant, and Alexa are just the beginning. Future apps will be capable of understanding and responding to user voice commands and conversations, enabling a hands-free and conversational user experience.

Progressive Web Apps (PWAs) combine the best of both web and mobile apps, offering the convenience of instant access and offline capabilities. These lightweight apps are poised to gain popularity, especially in regions with limited internet connectivity. Developers should consider PWAs as a versatile approach to app delivery.
As technology's impact on society becomes more pronounced, ethical considerations and sustainability will

play a vital role in app development. Developers will need to consider their apps' ethical implications, including data privacy, inclusivity, and environmental impact. Sustainable design principles will shape the way apps are developed and maintained.

The future of mobile app development will require a commitment to continuous learning and adaptation. Developers must stay agile, keep pace with emerging technologies, and be open to reimagining the possibilities of mobile apps. Learning from user feedback and adapting to changing user needs will be essential to building apps that stand the test of time.

In closing, the future of mobile app development is a realm of boundless innovation and transformative possibilities. As we reflect on the journey we've taken through this e-book, we recognize that the skills, insights, and best practices we've explored are not merely endpoints but stepping stones toward a future where mobile apps are more powerful, intuitive, and integral to our lives than ever before. Whether you're a seasoned developer or you're just beginning your journey, remember that you are part of a dynamic and influential community that shapes the digital landscape. Embrace the difficulties and opportunities that lie ahead, and with each line of code, you contribute to the ever-evolving story of mobile app development. The future is in your hands, and it's full of promise.

Thank you for buying and reading/ listening to our book. If you found this book useful/ helpful please take a few minutes and leave a review on the platform where you purchased our book. Your feedback matters greatly to us.